WEEKLY GEOGRAPHY PR

Table of Cont

Introduction	
Answer Key for Assessments	2
Additional Activities	3
Maps	4
Assessments	15
Unit 1: Community	25
Unit 2: State	43
Unit 3: Region	59
Unit 4: Country	71
Unit 5: World	83

Introduction

The modern world has woven a web of interdependency. Each part of the world gives to and takes from the other parts. We gain a better sense of where and who we are today through a knowledge of geography. By giving students geographical information and allowing them to use their own experiences, we can help them to connect the familiar in their lives with the unfamiliar. They can then move more easily from the well-known features of locale and state to the less familiar concepts of region, country, and world. Students can begin with the area they know and then expand their knowledge base. By giving students a better knowledge of geography, we give them a better understanding of the world in which they live. Such knowledge will help students in their standardized testing and in their broader academic pursuits.

General standards in geography for this grade suggest that students should be able to perform a variety of skills. Students should be able to understand and use map components, employ the cardinal and intermediate directions, use latitude and longitude, identify the continents and oceans, and recognize landforms. They should understand regional similarities and differences, distinguish between physical and human features, and be able to discuss the relationship between humans and the environment. This book adheres to these standards.

Organization
The book provides a collection of 36 weeks of geography practice activities that highlight a different theme of geography each day of the week. On the back of each weekly practice student page, there is a corresponding teacher page with answers and additional information.

The book is organized using a logical movement from the most specific and familiar aspect of geography—community—to the most general and most unfamiliar—Earth as a part of the solar system. This organization allows the students to understand better the increasing rings of relationship in their world.

Use
Weekly Geography Practice is meant to supplement the social studies curriculum in the classroom. Map skills help students to improve their sense of location, place, and movement. This book is designed as a flexible tool for the teacher. The practice lessons may be used in a wide variety of settings and modes. The weekly lessons may be used in sequence, or you may prefer to use the lessons wherever they best fit into your curriculum plans. You may photocopy the entire page of activities and hand them out to students at the beginning of the week, cut the pages into strips and pass them out daily, maintain the sheets in a folder to use at a specific time each day, assign the activities as homework, or use them as overhead transparencies.

The weekly lesson pages may be used for any of the following activities:

- warm-ups for the social studies class,
- cooperative group activities,
- independent student-monitored practices at a learning station,
- make-up assignments,
- additional assignments for students with specific needs,
- extra-credit assignments,
- or as short transitional activities between periods of longer concentration.

The Five Themes of Geography
An effective approach to the study of geography can be organized around the five themes of geography. Each day of the week is dedicated to a particular theme of geography, which is represented with a graphic icon to serve as a visual cue and reminder of the theme. The weekly practice questions are divided among these five themes as shown:

Monday — LOCATION

Where is it? Location tells where a place is located. It can include latitude and longitude, cardinal or intermediate directions, or even just the words "next to."

Tuesday — PLACE

What does it look like? What are its surroundings? Place is the part of an area you see. It includes landforms, rivers, seats of government, or even buildings in a city.

Wednesday — MOVEMENT

How do people, goods, and ideas move from one place to another? Movement looks at how and why things move. It looks at why people leave a country, how goods are moved, and how both affect the land or people.

Thursday — REGION

How are different places in an area alike? Region looks at the way an area is divided. A region has characteristics that are the same. Often areas are grouped because the people share a language, people do the same kind of work, or the climate or landforms are the same.

Friday — INTERACTION

How do people use or change the land? Interaction is how people use or change the environment. It considers where people have parks, build cities, set up factories, or farm.

Assessments
Each of the five units in the book includes a pre-assessment test and a post-assessment test. As suggested, one assessment is intended to be completed before the unit of study is begun. This assessment gauges the students' basic knowledge of the unit's concept. The other assessment is intended to be completed when the unit of study is finished. This assessment gauges how well the students have learned the concepts presented in the unit.

Maps and Globes
Students gain a greater sense of place by knowing their relationship to other places. For this reason, maps and globes are important tools. This book includes 20 reproducible maps for use with the weekly practice sheets. Maps are also included for the assessment tests. At least one weekly practice sheet requires the use of a globe. In any case, various maps, especially of the United States and the world, are a handy addition to any classroom.

Additional Activities
Most of the weekly practice questions require short answers. Some questions are more substantial and require a longer answer. You might consider assigning these as homework. In addition, on page 3 is a series of additional activities tailored to each unit. These can be used as extra-credit assignments, research projects, or group work.

Notes
1. Each daily practice lesson should take no more than five to ten minutes for an average student.
2. Students may complete the lessons on their own paper, if necessary.
3. Several of the maps will be used more than once, so students should be told to retain their copies.
4. Some weekly practice series suggest the use of more than one map. In these cases, the students are often expected to gather information from at least two maps and put the information together to produce the answer.

Answer Key for Assessments
page 15: 1. C, 2. C, 3. B, 4. B, 5. A, 6. D
page 16: 1. B, 2. C, 3. D, 4. A, 5. D, 6. D
page 17: 1. C, 2. D, 3. B, 4. B, 5. D, 6. C
page 18: 1. B, 2. B, 3. D, 4. C, 5. A, 6. C
page 19: 1. C, 2. B, 3. B, 4. A, 5. D, 6. C
page 20: 1. D, 2. B, 3. D, 4. A, 5. A, 6. B
page 21: 1. B, 2. C, 3. D, 4. D, 5. A, 6. C
page 22: 1. C, 2. B, 3. A, 4. D, 5. B, 6. C
page 23: 1. B, 2. A, 3. C, 4. C, 5. C, 6. D
page 24: 1. C, 2. B, 3. D, 4. A, 5. C, 6. B

Additional Activities

Unit 1
1. Is there a special building in your community? Do research to find out about its history. Then, write a short report about the building. Include a drawing of the building with your report.

2. Write a short story about the founding of your community. Do you know when your community was founded? Do you know about the people who founded it? Were there Native American groups in the area? If you do not know exactly, you can make up parts of your story. Pretend you are one of the founders. Why would you choose this place for your community?

3. Plan a campaign for traffic safety. Have the students plan and draw posters that promote pedestrian safety, crosswalk safety, bike safety, and bus and car safety. Display the completed posters in the school.

Unit 2
1. Work in a group of four or five students. Learn the state song. Sing it for your class.

2. Find out about a famous person from your state. Do research to find out more about that person. Then, write a short biography of that person.

3. Work in a group of four or five students. Write a short play about some important event in the history of your state. Have each person in the group play a role. Write dialogue for the play. When the play is finished, act it out for the class.

4. Find a folk tale about your state. With a group of classmates, act out the folk tale in class.

Unit 3
1. Using modeling clay, make a relief map of your region. Show mountains, hills, valleys, plains, rivers, and lakes. Scratch lines into the clay to indicate the state borders. Display your completed map in the classroom.

2. To demonstrate the difficulty of mapmaking without the aid of overhead views or sophisticated instruments, have the students make a map of the schoolyard and building. If they are more ambitious, they can make a map of their neighborhood or community. Discuss the difficulties of their mapmaking experience. Display the completed maps in the classroom.

3. Find out about an environmental problem in the region. Make a poster that tells about the problem. Include some pictures. How can the problem be fixed? Display the completed poster in the classroom.

Unit 4
1. Find the lines of latitude and longitude near your town. Use the lines to estimate your town's location. When you write latitude and longitude, you write the line of latitude first. For example, the latitude and longitude for Milwaukee, Wisconsin, is about 43° N, 88° W. Write the latitude and longitude of your town.

2. Follow the line of latitude near your town east or west. Find a big city somewhere in the world that is near that line. Write the name of the city, its country, and its continent. Now, follow the line of longitude near your town north or south. Find a city somewhere in the world near that line. Write the name of the city, its country, and its continent.

3. Get your teacher or an adult to help you with this activity. Write down the latitude and longtitude for your town. Then, go on the Internet.
Go to this address:
http://www.fourmilab.ch/cgi-bin/uncgi/Earth
This is the web address for EarthCam. At this site, you can type in your latitude and longitude. Then, you can see your area from a camera in space!

Unit 5
1. Do research to find out about the first explorers to reach the North Pole and the South Pole. Then, write a short report or a short story about one of the explorers who traveled to the pole.

2. Choose one of the other planets in the solar system. Do research to find out more about that planet. Then, write a report about your planet. Include a drawing of your planet with your report.

3. Write a short story about a trip into space. What would it be like to be the first human visitor to Mars or some other planet?

Name _____ Date _____

Map A

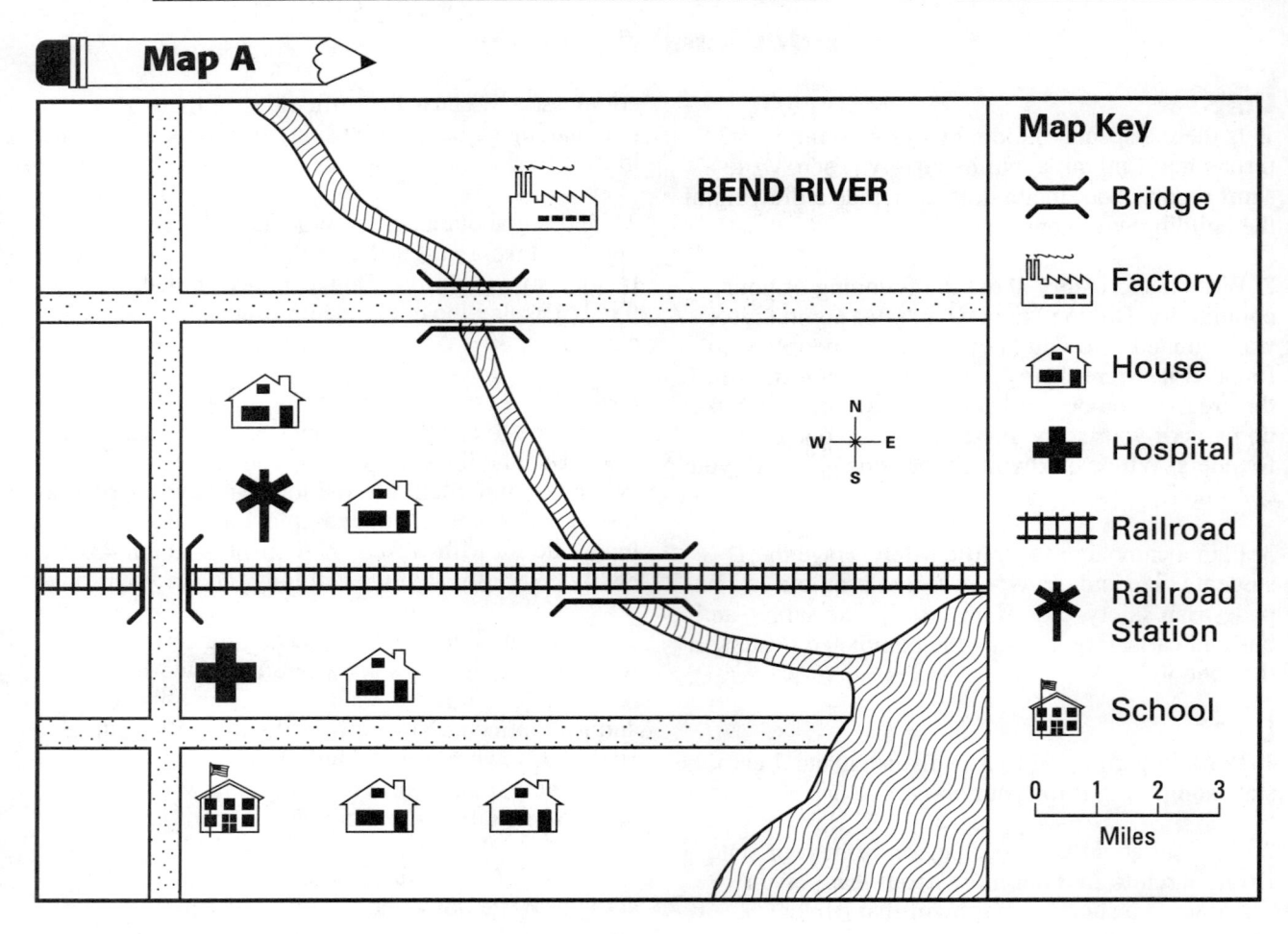

BEND RIVER

Map Key
- Bridge
- Factory
- House
- Hospital
- Railroad
- Railroad Station
- School

0 1 2 3
Miles

Map B

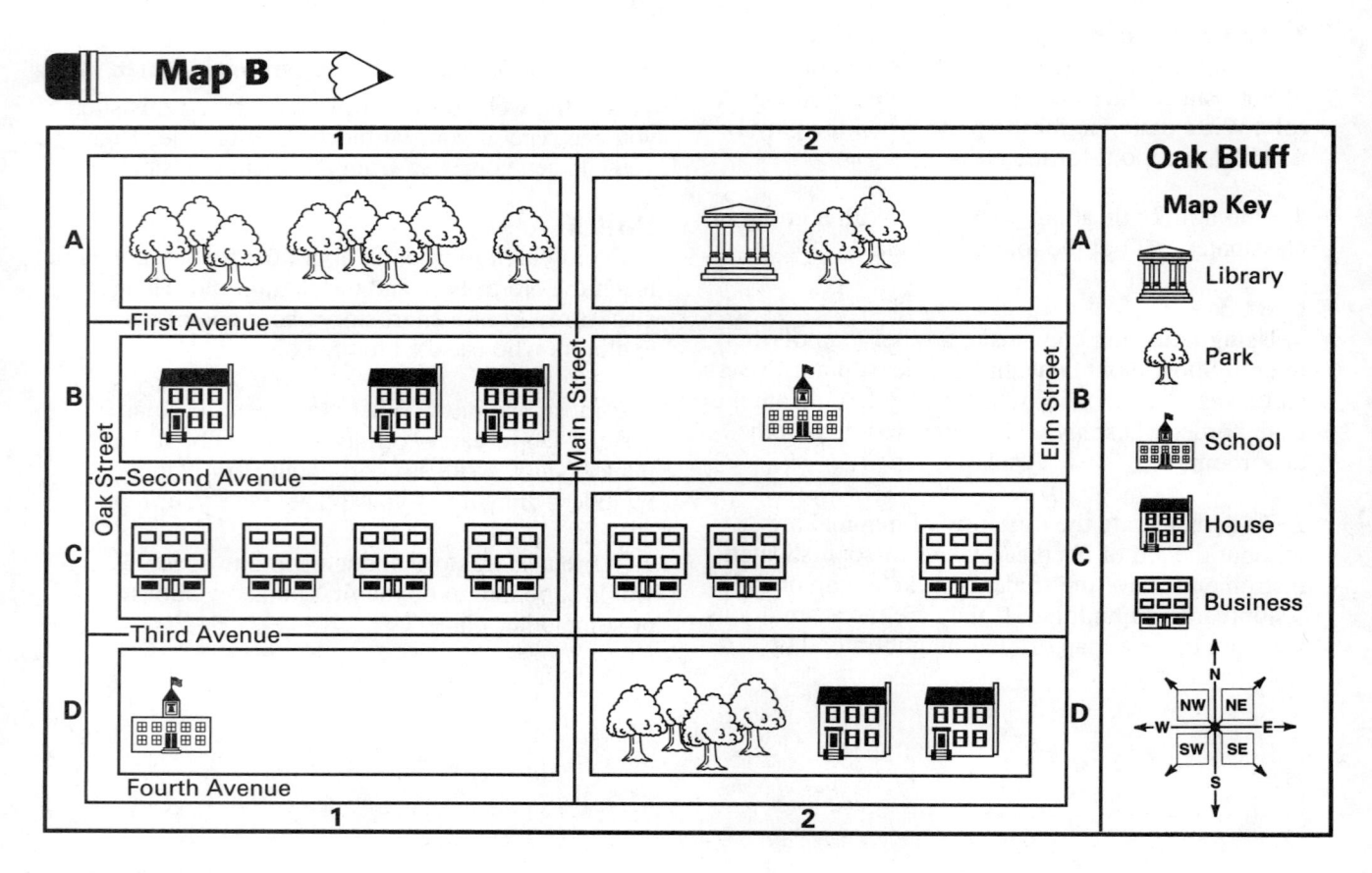

Oak Bluff

Map Key
- Library
- Park
- School
- House
- Business

First Avenue
Second Avenue
Third Avenue
Fourth Avenue

Oak Street
Main Street
Elm Street

www.svschoolsupply.com
© Steck-Vaughn Company

4

Maps
Weekly Geography Practice 3, SV 3432-0

Name _____ Date _____

Map C

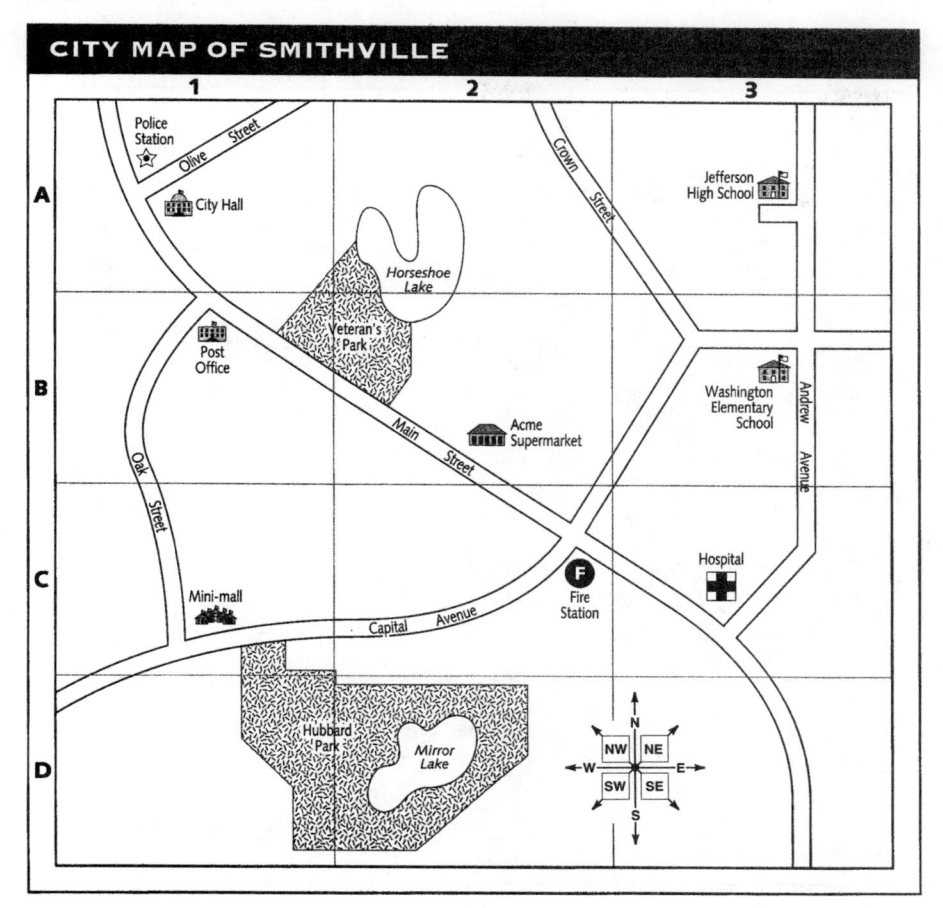

CITY MAP OF SMITHVILLE

Map D

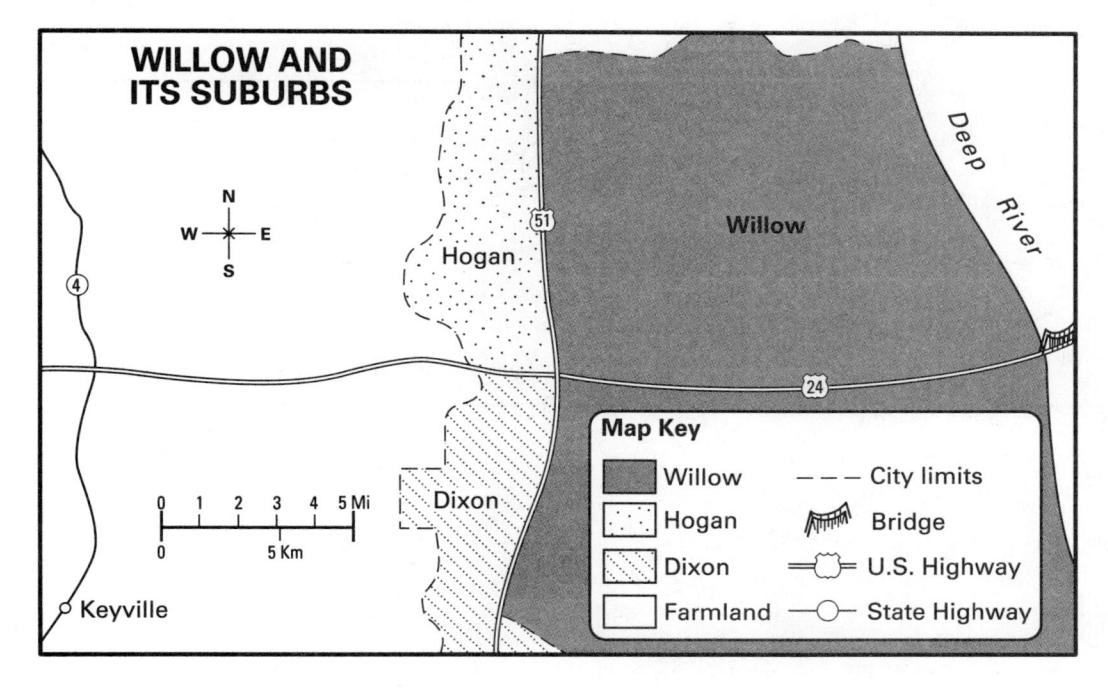

WILLOW AND ITS SUBURBS

Map Key
- Willow
- Hogan
- Dixon
- Farmland
- – – – City limits
- Bridge
- U.S. Highway
- State Highway

Name _____ Date _____

Map E

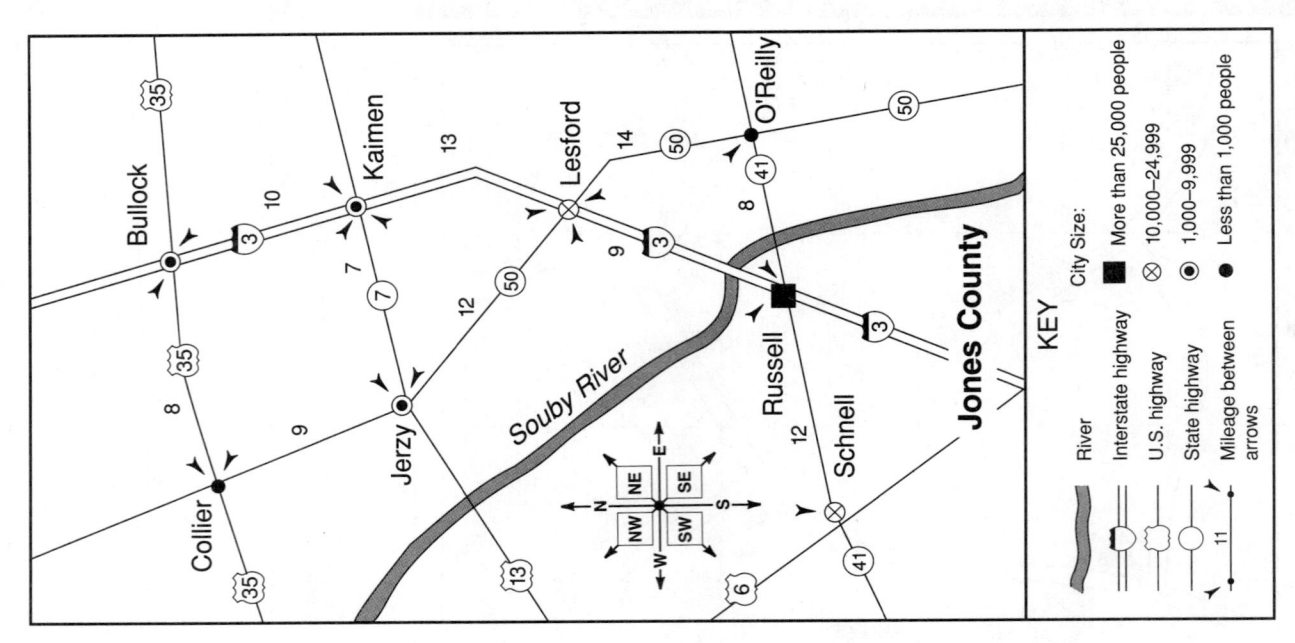

Map F

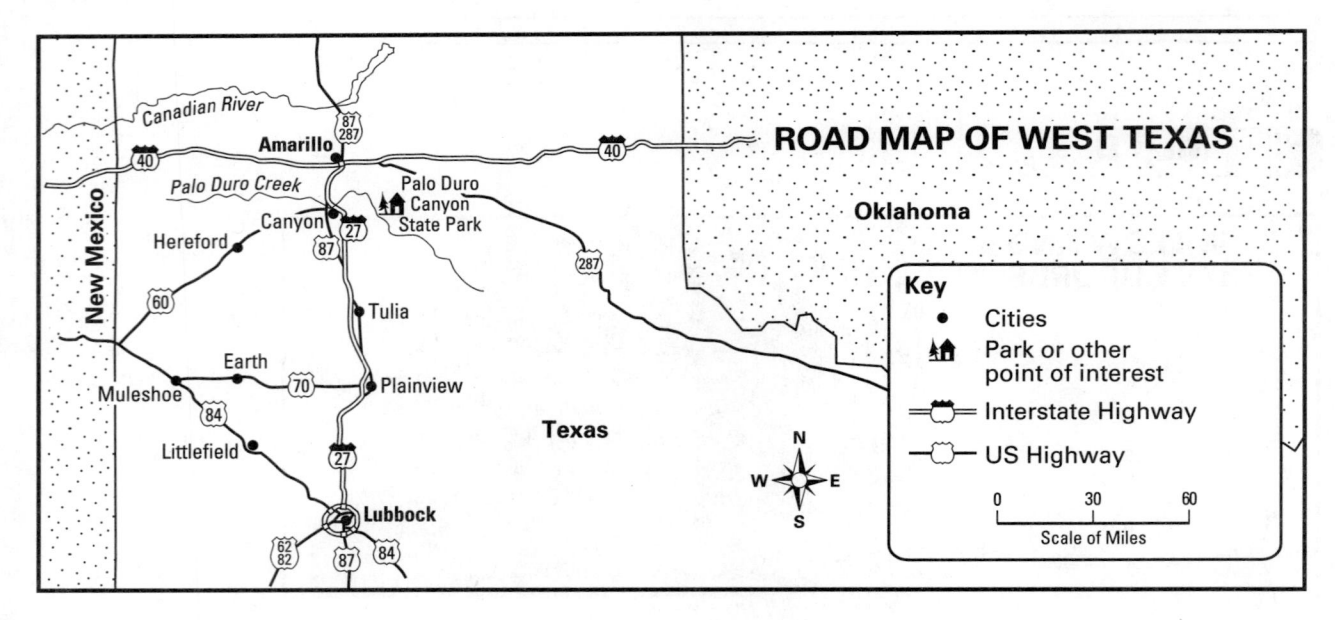

ROAD MAP OF WEST TEXAS

Name _____ Date _____

Map G

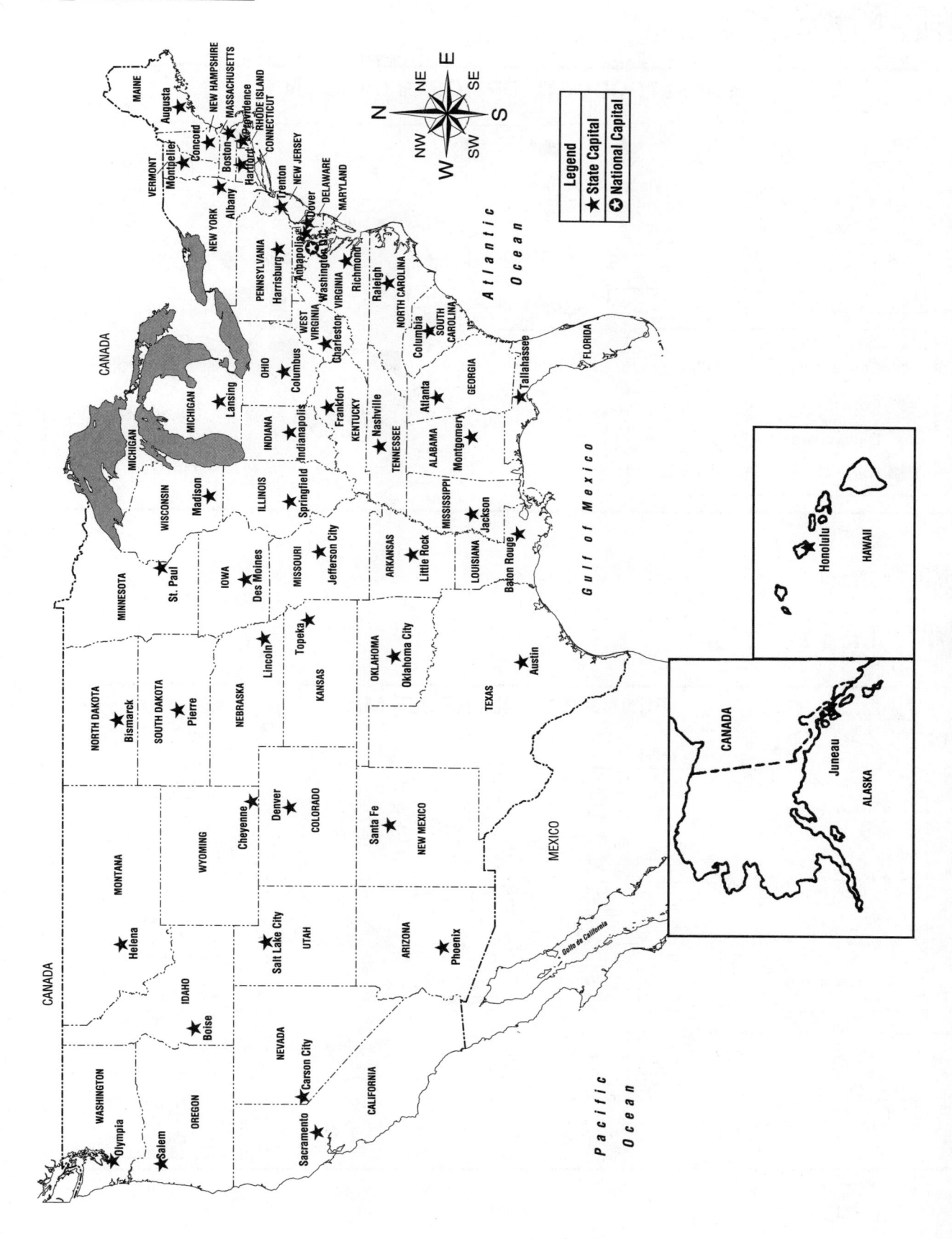

Name _____ Date _____

Map H

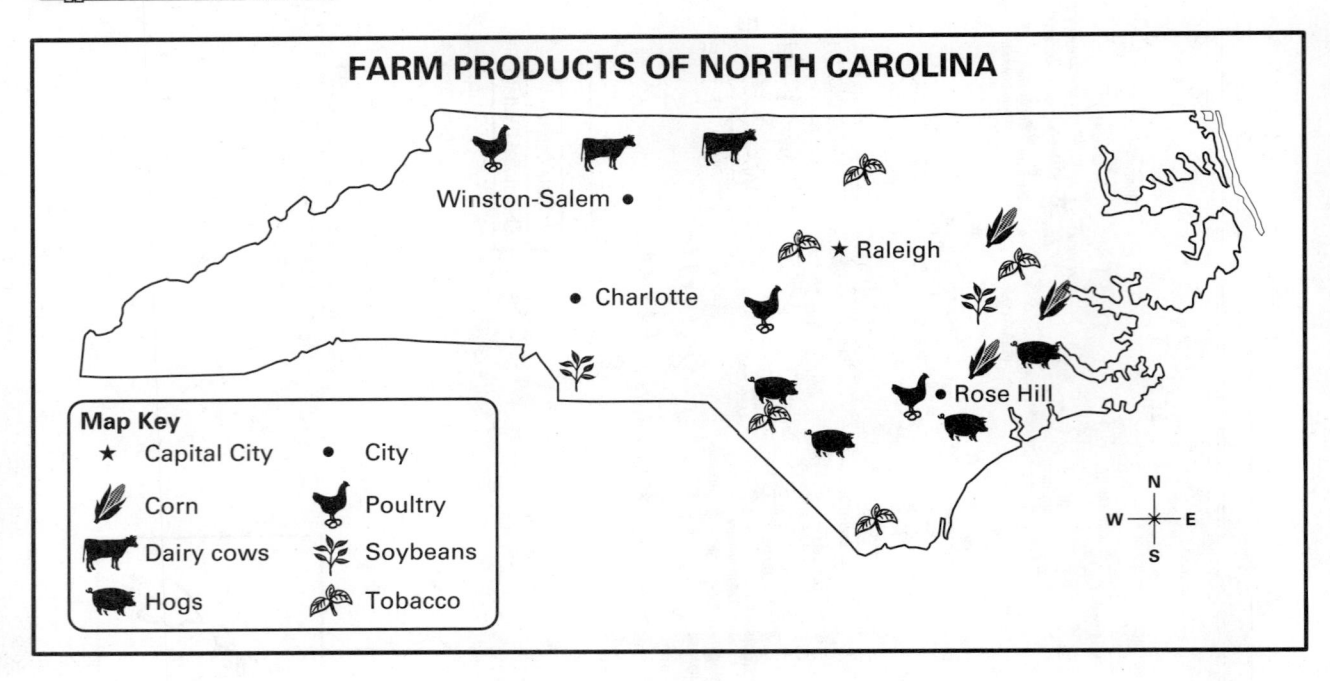

FARM PRODUCTS OF NORTH CAROLINA

Map Key
- ★ Capital City
- • City
- Corn
- Poultry
- Dairy cows
- Soybeans
- Hogs
- Tobacco

Map I

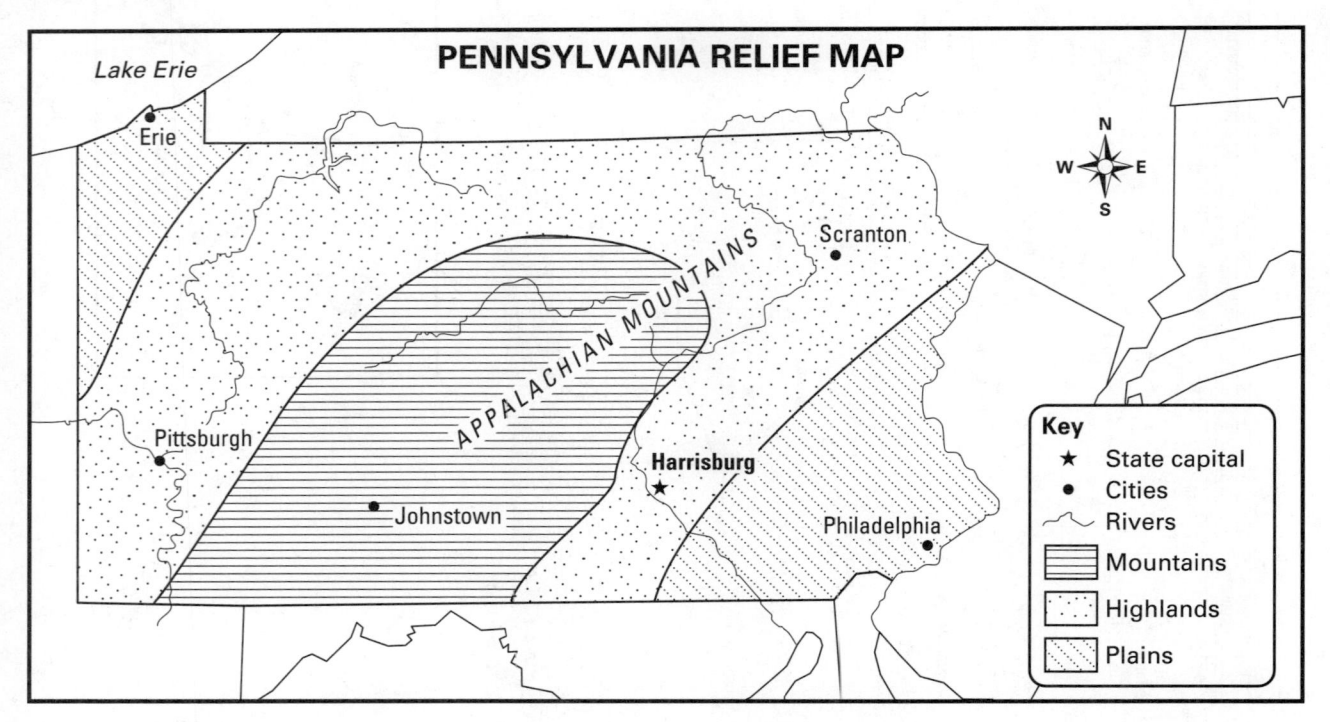

PENNSYLVANIA RELIEF MAP

Key
- ★ State capital
- • Cities
- Rivers
- Mountains
- Highlands
- Plains

Name _____ Date _____

Map J

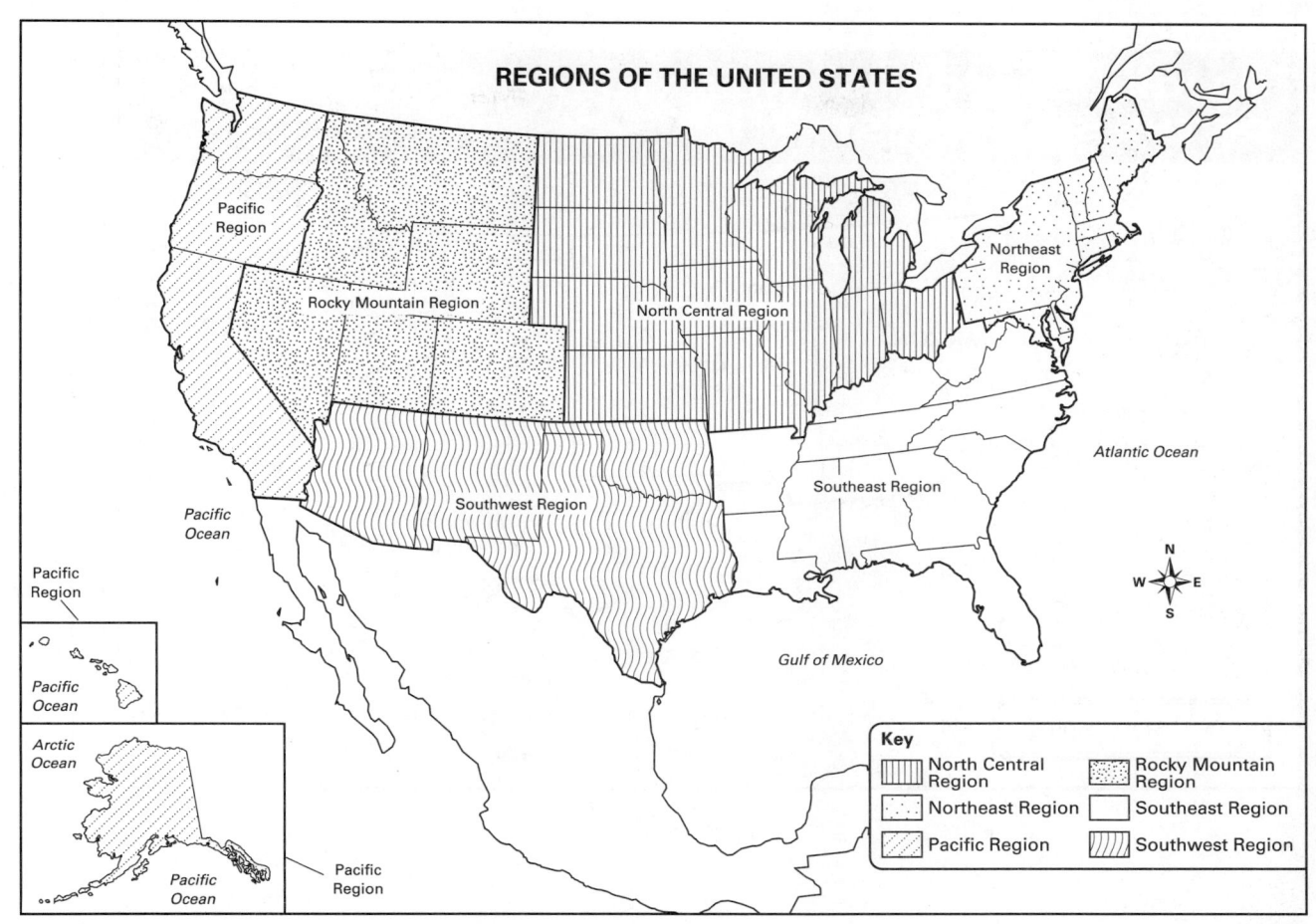

REGIONS OF THE UNITED STATES

Key
- North Central Region
- Northeast Region
- Pacific Region
- Rocky Mountain Region
- Southeast Region
- Southwest Region

Map K

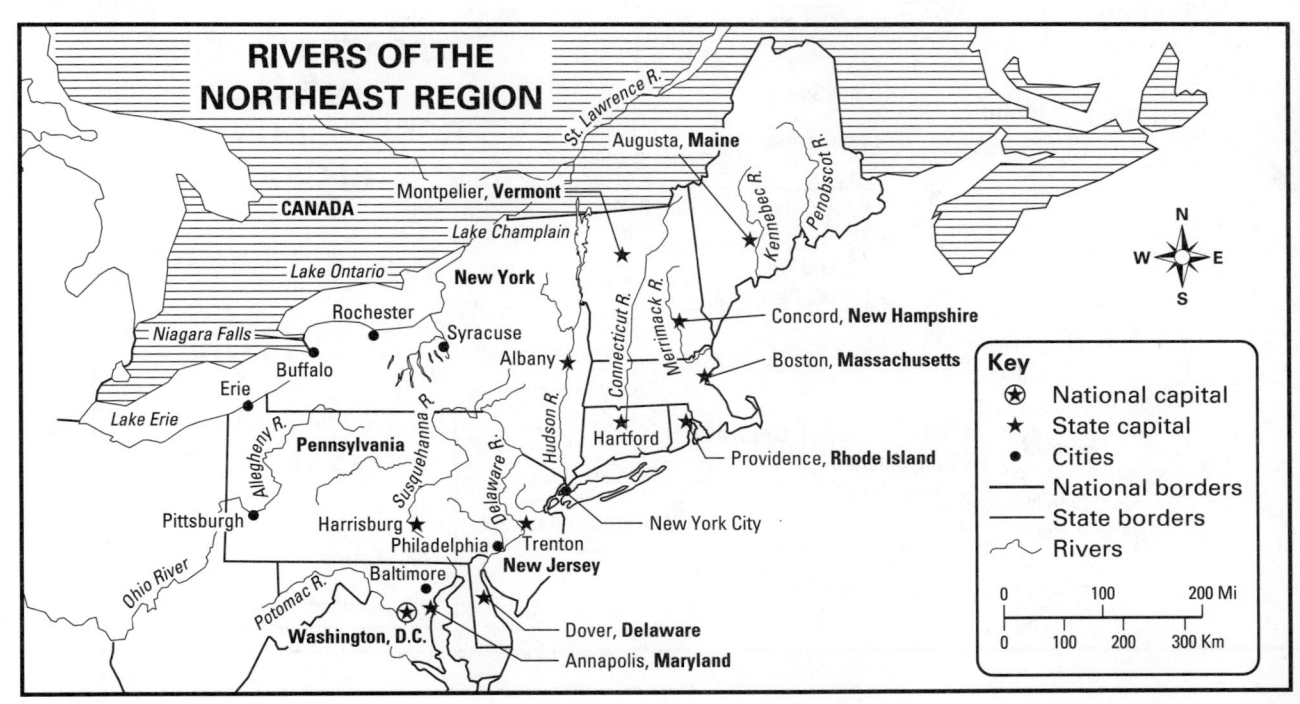

RIVERS OF THE NORTHEAST REGION

Key
- ⊛ National capital
- ★ State capital
- • Cities
- National borders
- State borders
- Rivers

0   100   200 Mi
0   100   200   300 Km

Name _____ Date _____

Map L

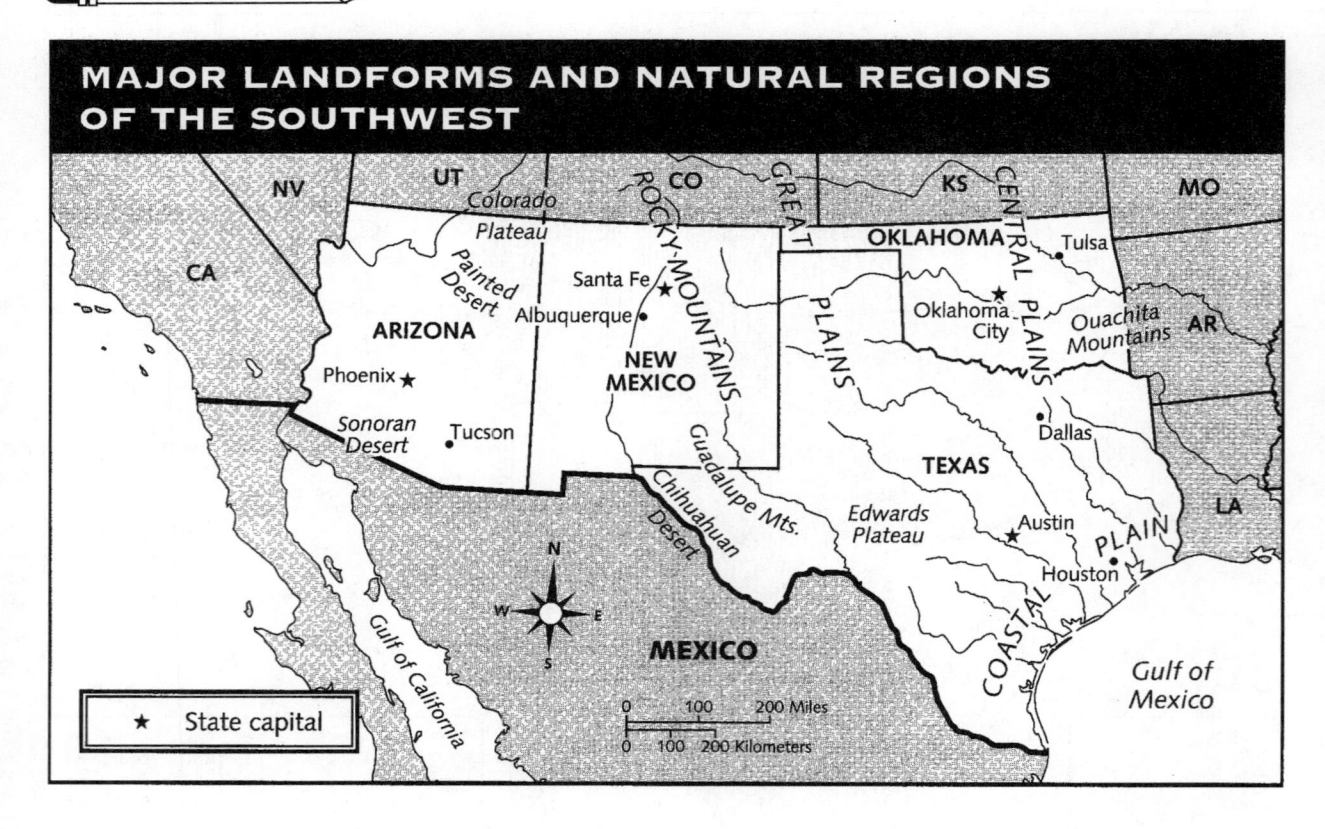

MAJOR LANDFORMS AND NATURAL REGIONS OF THE SOUTHWEST

Map M

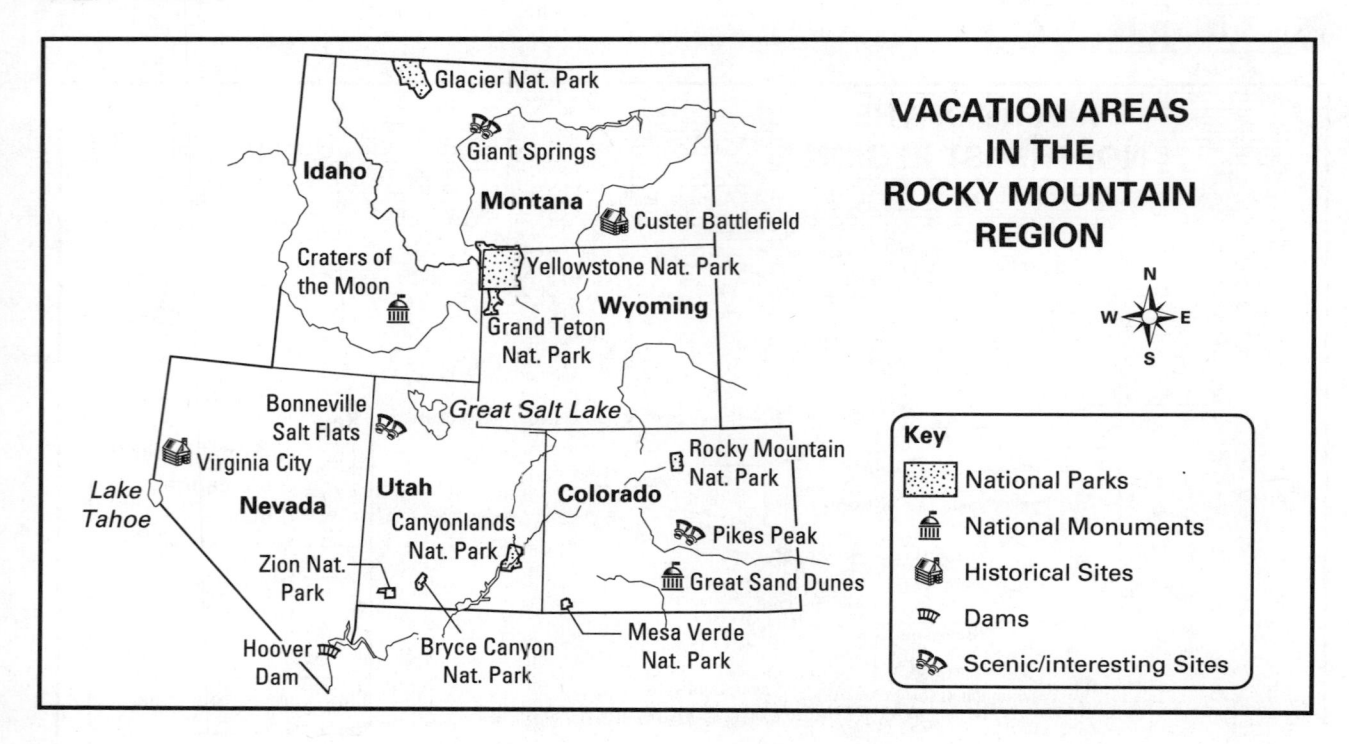

VACATION AREAS IN THE ROCKY MOUNTAIN REGION

Name _____ Date _____

Map N

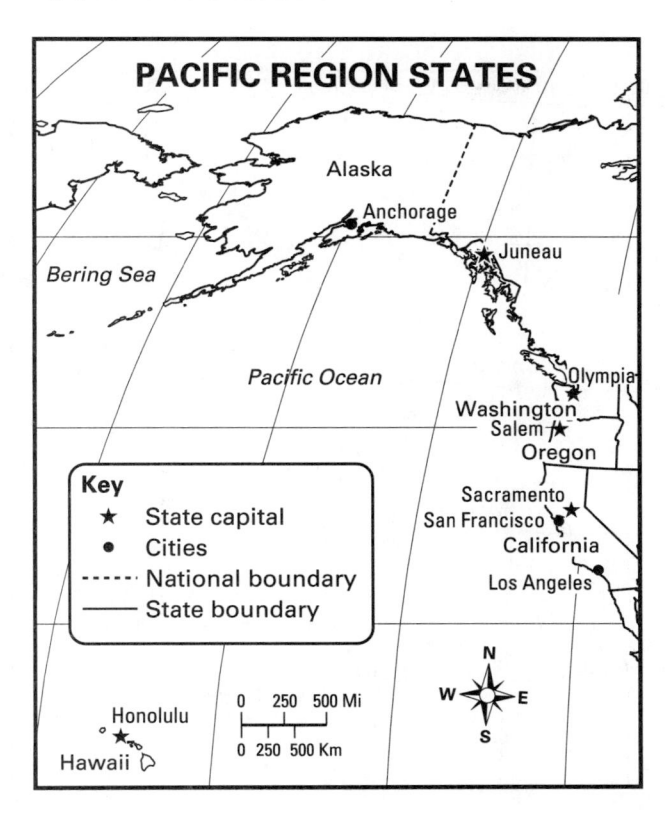

PACIFIC REGION STATES

- Alaska
- Anchorage
- Juneau
- Bering Sea
- Pacific Ocean
- Olympia
- Washington
- Salem
- Oregon
- Sacramento
- San Francisco
- California
- Los Angeles
- Honolulu
- Hawaii

Key
- ★ State capital
- • Cities
- ---- National boundary
- —— State boundary

0 250 500 Mi
0 250 500 Km

Map O

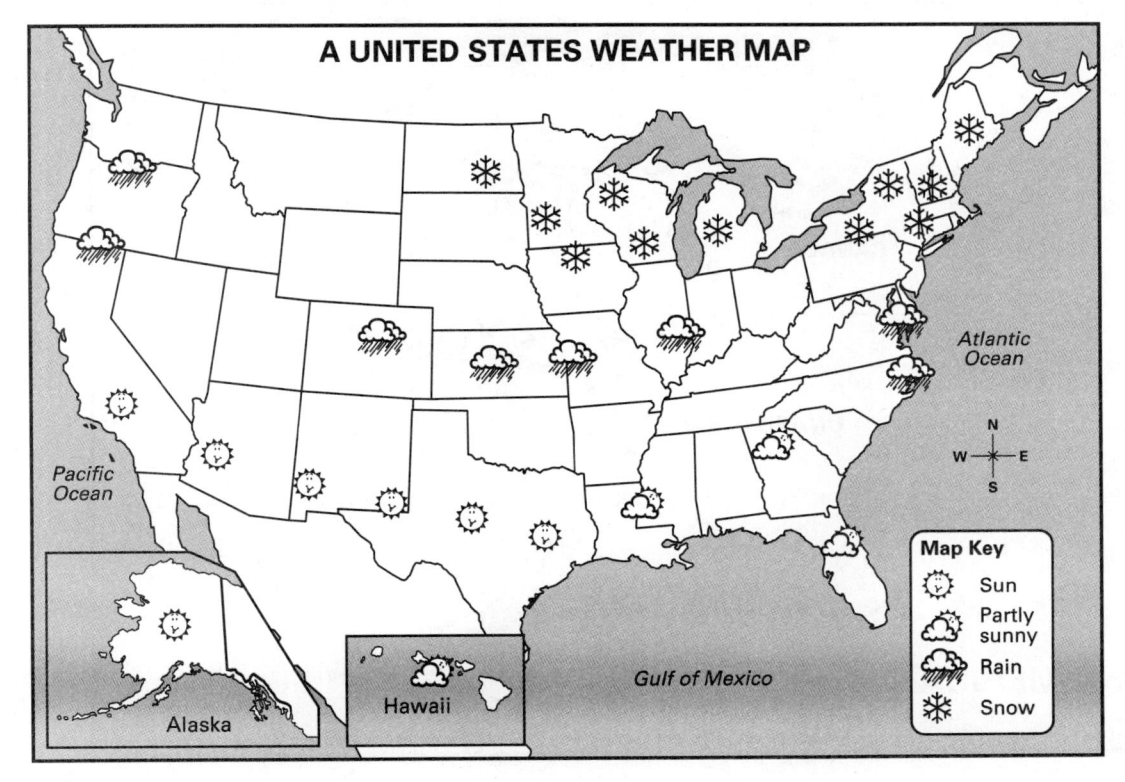

A UNITED STATES WEATHER MAP

Atlantic Ocean
Pacific Ocean
Alaska
Hawaii
Gulf of Mexico

Map Key
- Sun
- Partly sunny
- Rain
- Snow

www.svschoolsupply.com
© Steck-Vaughn Company

11

Maps
Weekly Geography Practice 3, SV 3432-0

Name _____ Date _____

Map P

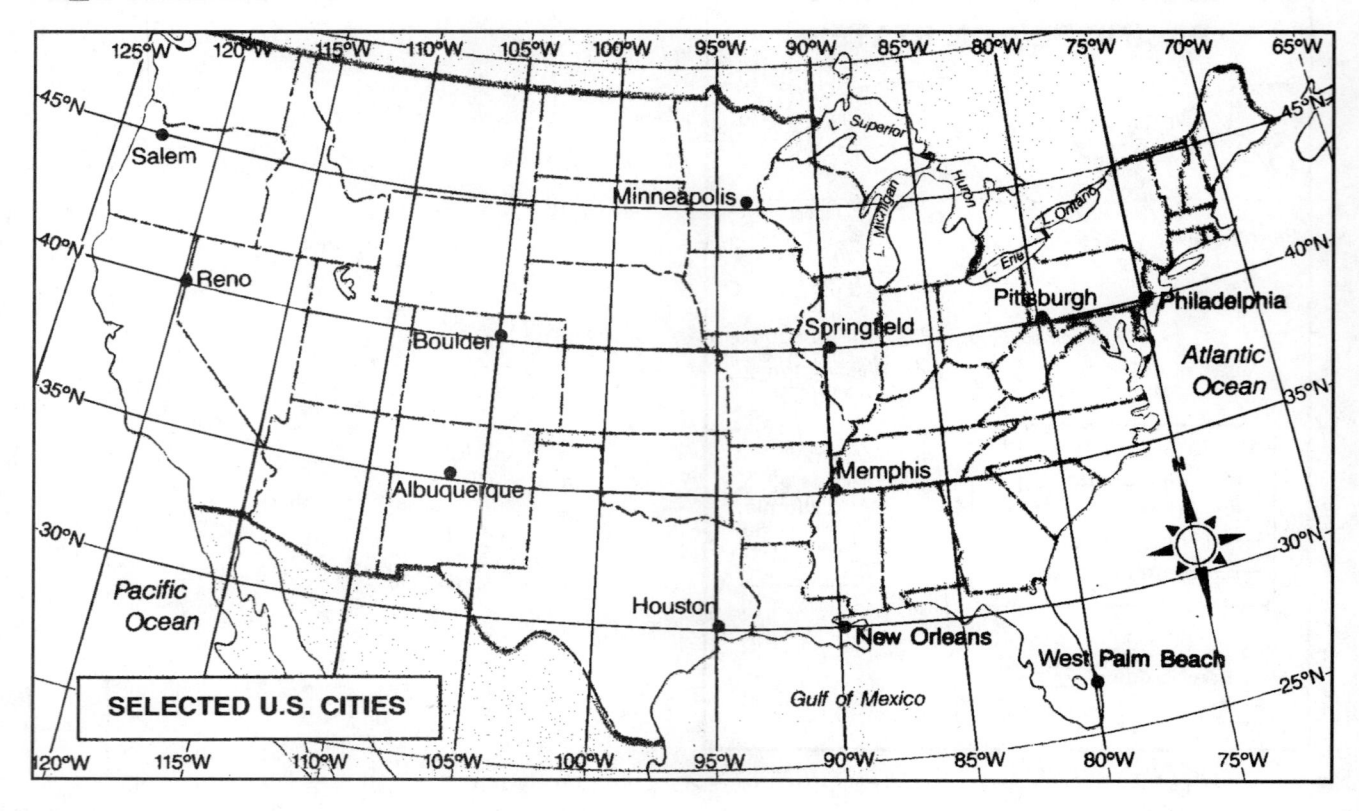

SELECTED U.S. CITIES

Map Q

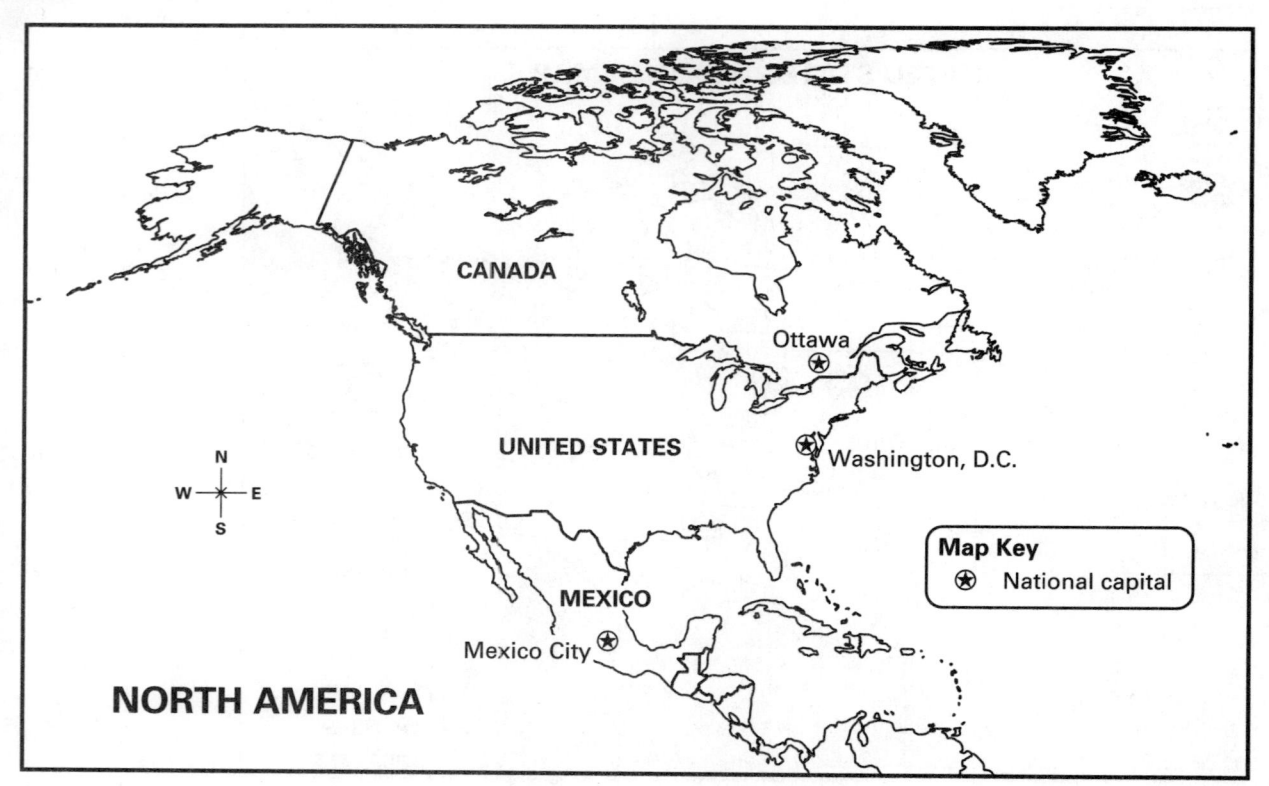

NORTH AMERICA

Map Key
★ National capital

Name _____ Date _____

Map R

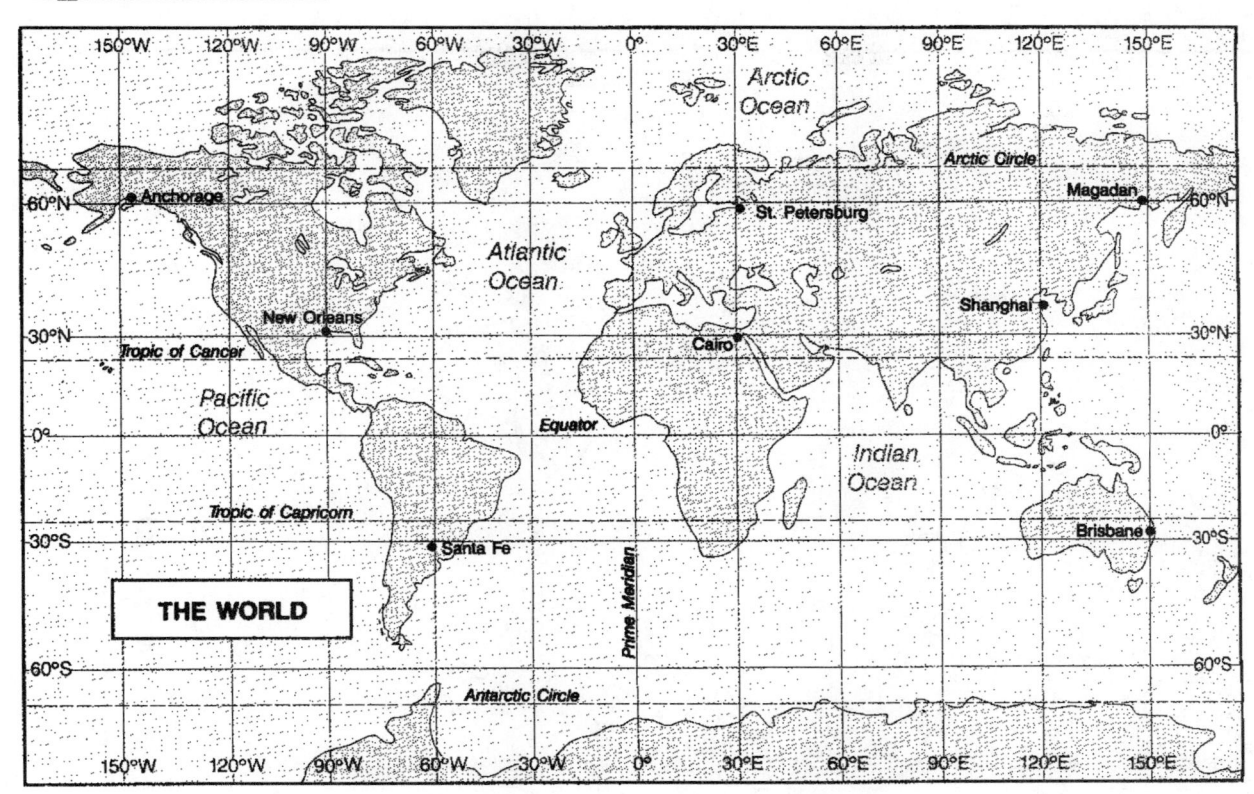

THE WORLD

Map S

Mercury Venus Earth Mars Jupiter Saturn Uranus Neptune Pluto

Name _____ Date _____

Map T

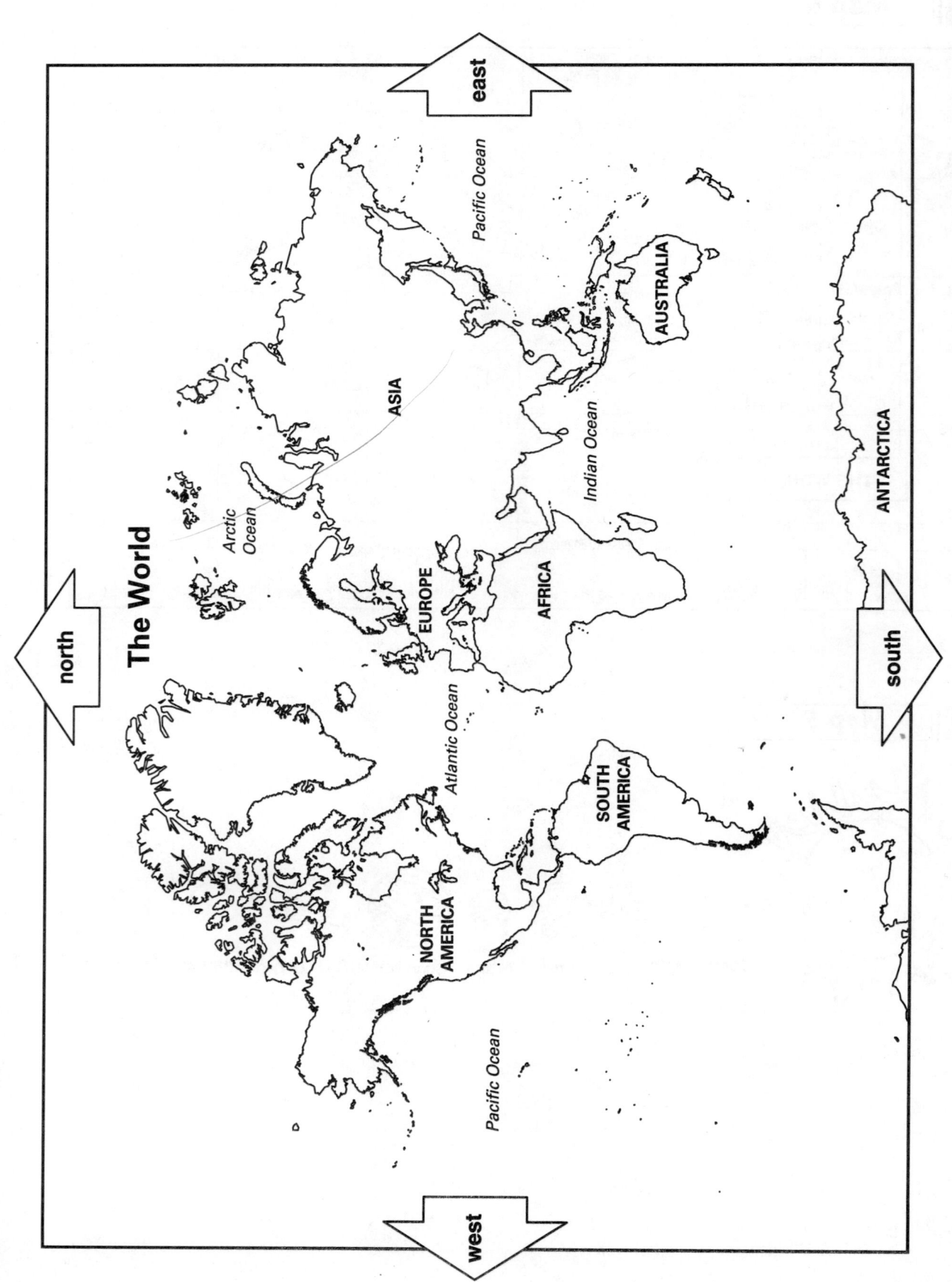

The World

east

north

south

west

- Pacific Ocean
- ASIA
- AUSTRALIA
- Arctic Ocean
- Indian Ocean
- EUROPE
- AFRICA
- ANTARCTICA
- Atlantic Ocean
- NORTH AMERICA
- SOUTH AMERICA
- Pacific Ocean

www.svschoolsupply.com
© Steck-Vaughn Company

Maps
Weekly Geography Practice 3, SV 3432-0

Name _Caleb_   Date _Oct. 2003_

UNIT 1 PRE-ASSESSMENT

Directions: Use the map to answer the questions. Darken the circle by the correct answer to each question.

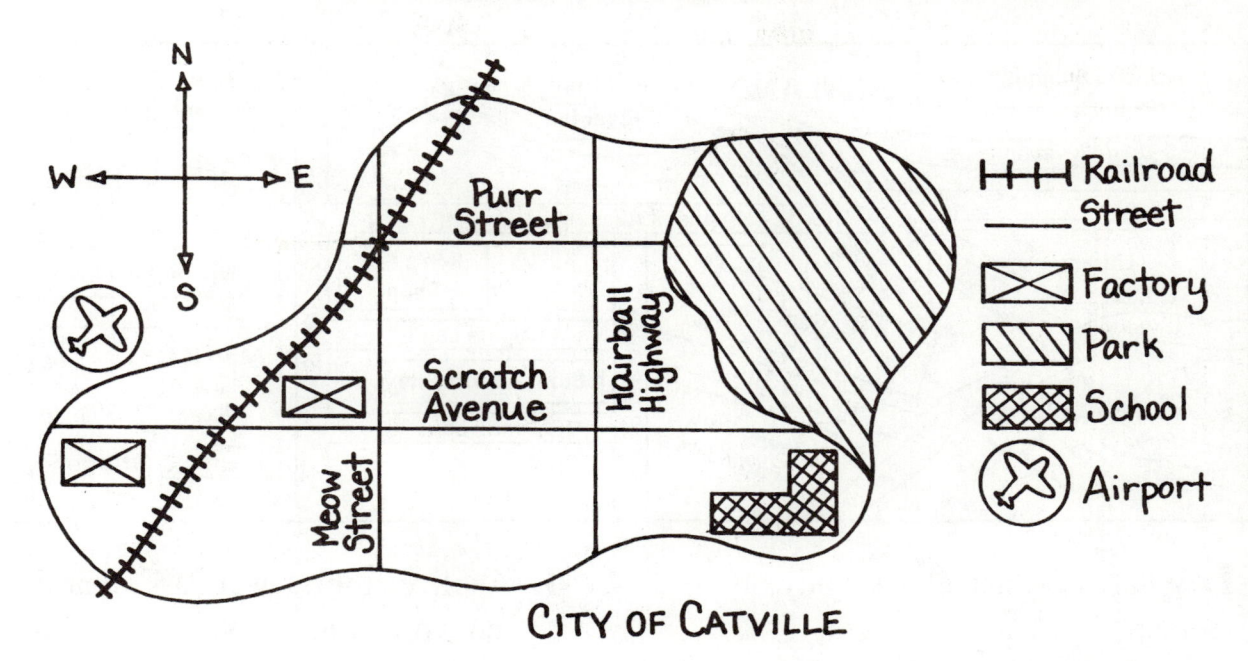

CITY OF CATVILLE

1. What is the name of the city on the map?
   - Ⓐ Scratch Avenue
   - Ⓑ Airport
   - ●Ⓒ Catville
   - Ⓓ Purr Street

2. How many factories are in this city?
   - Ⓐ none
   - Ⓑ one
   - ●Ⓒ two
   - Ⓓ three

3. Which two streets will take you to the park?
   - Ⓐ Purr Street and Hairball Highway
   - ●Ⓑ Purr Street and Scratch Avenue
   - ●Ⓒ Meow Street and Scratch Avenue
   - Ⓓ Meow Street and Hairball Highway

4. The railroad runs through the intersection of which two streets?
   - Ⓐ Purr Street and Hairball Highway
   - Ⓑ Meow Street and Purr Street
   - Ⓒ Purr Street and Scratch Avenue
   - Ⓓ Meow Street and Scratch Avenue

5. In which direction would you travel to get from the school to the park?
   - ●Ⓐ north
   - Ⓑ south
   - Ⓒ east
   - Ⓓ west

6. In which direction would you travel to get from the park to the airport?
   - Ⓐ north
   - Ⓑ south
   - Ⓒ east
   - ●Ⓓ west

Name _____ Date _____

UNIT 1 POST-ASSESSMENT

Directions: Use the map to answer the questions. Darken the circle by the correct answer to each question.

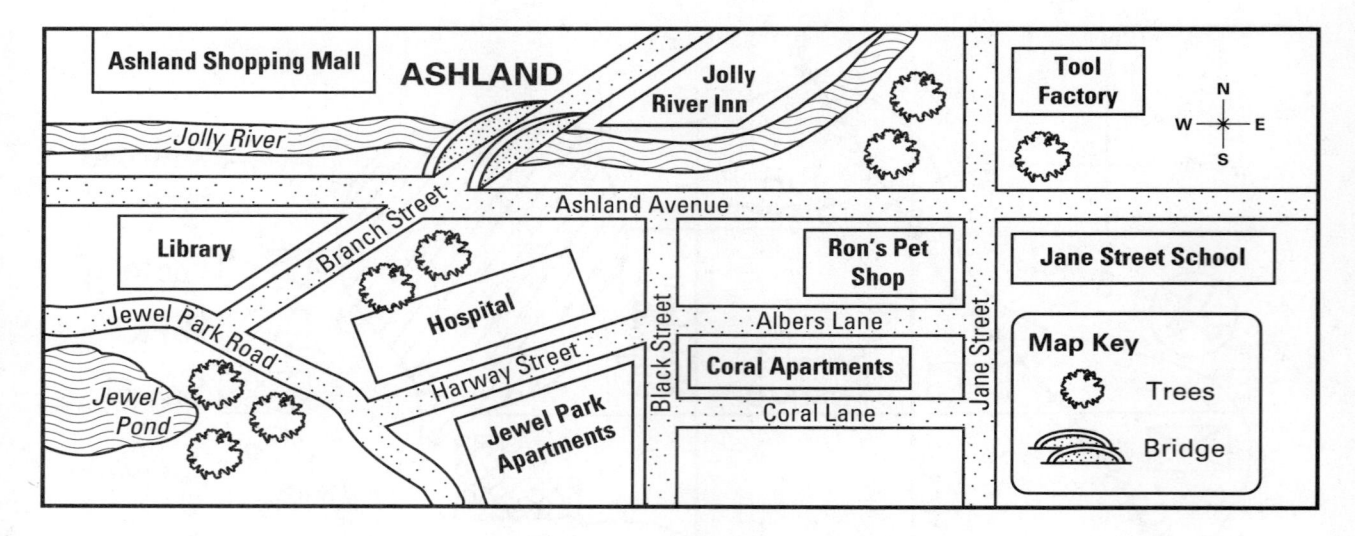

1. What is the name of the city on the map?
   - Ⓐ Jewel Park
   - Ⓑ Ashland
   - Ⓒ Jolly River
   - Ⓓ Tool Factory

2. Which street crosses Jolly River?
   - Ⓐ Ashland Avenue
   - Ⓑ Jewel Park Road
   - Ⓒ Branch Street
   - Ⓓ Coral Lane

3. Which street runs by Jewel Pond?
   - Ⓐ Black Street
   - Ⓑ Coral Avenue
   - Ⓒ Ashland Avenue
   - Ⓓ Jewel Park Road

4. You live at the Jewel Park Apartments and work at the Tool Factory. Which direction is your job from your home?
   - Ⓐ northeast
   - Ⓑ southeast
   - Ⓒ northwest
   - Ⓓ southwest

5. Which direction is the Ashland Shopping Mall from the Jane Street School?
   - Ⓐ northeast
   - Ⓑ southeast
   - Ⓒ southwest
   - Ⓓ northwest

6. You walk from the Jolly River Inn to the Jewel Park Apartments. Which place would you pass taking Branch Street to Jewel Park Road?
   - Ⓐ Jane Street School
   - Ⓑ Ron's Pet Shop
   - Ⓒ Coral Apartments
   - Ⓓ Library

www.svschoolsupply.com
© Steck-Vaughn Company

Assessments
Weekly Geography Practice 3, SV 3432-0

Name _____ Date _____

UNIT 2 PRE-ASSESSMENT

Directions: Use the map to answer the questions. Darken the circle by the correct answer to each question.

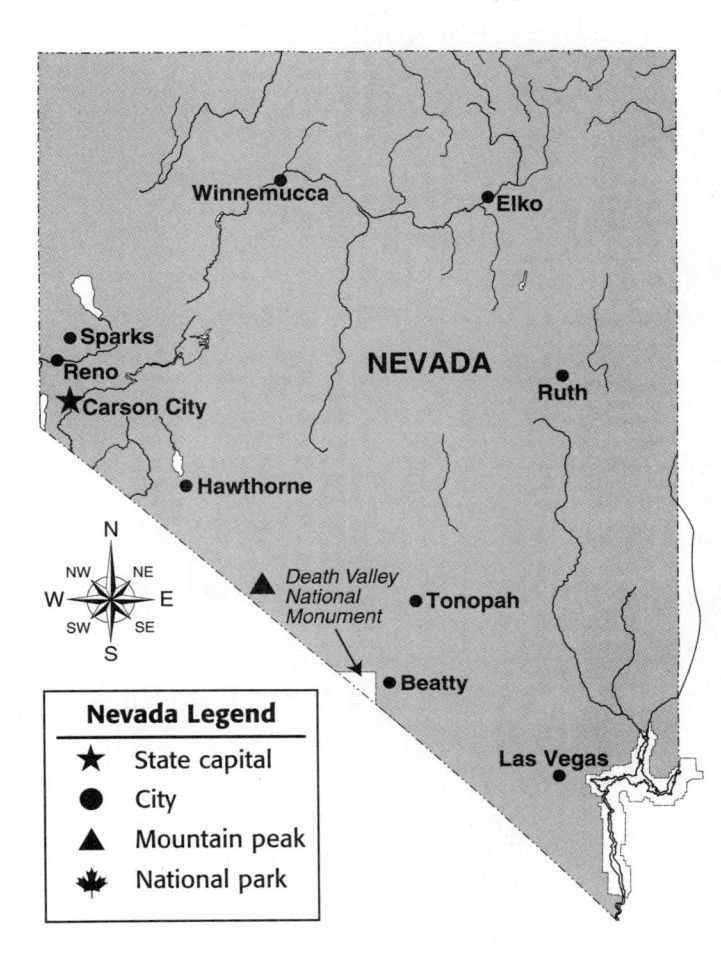

Nevada Legend
- ★ State capital
- ● City
- ▲ Mountain peak
- ❋ National park

1. What state is shown on the map?
   - Ⓐ Carson City
   - Ⓑ Death Valley
   - Ⓒ Nevada
   - Ⓓ Legend

2. What is the capital city of this state?
   - Ⓐ Las Vegas
   - Ⓑ Ruth
   - Ⓒ Winnemucca
   - Ⓓ Carson City

3. What famous place is located near Beatty?
   - Ⓐ Mount Rushmore
   - Ⓑ Death Valley National Monument
   - Ⓒ Yellowstone National Park
   - Ⓓ Disneyland

4. To travel from Sparks to Carson City, which direction must you go?
   - Ⓐ north
   - Ⓑ south
   - Ⓒ east
   - Ⓓ west

5. To travel from Ruth to Reno, which direction must you go?
   - Ⓐ north
   - Ⓑ south
   - Ⓒ east
   - Ⓓ west

6. To travel from Hawthorne to Tonopah, which direction must you go?
   - Ⓐ northeast
   - Ⓑ southwest
   - Ⓒ southeast
   - Ⓓ northwest

Name _____ Date _____

UNIT 2 POST-ASSESSMENT

Directions → Use the map to answer the questions. Darken the circle by the correct answer to each question.

1. Which state borders South Dakota to the south?
   - Ⓐ Montana
   - Ⓑ Nebraska
   - Ⓒ Minnesota
   - Ⓓ North Dakota

2. Which state borders South Dakota to the southeast?
   - Ⓐ Montana
   - Ⓑ Iowa
   - Ⓒ North Dakota
   - Ⓓ Nebraska

3. If you wanted to fly from Aberdeen to Rapid City, which direction would you go?
   - Ⓐ east
   - Ⓑ southeast
   - Ⓒ northeast
   - Ⓓ southwest

4. About how far is Pierre from Rapid City?
   - Ⓐ 50 miles
   - Ⓑ 100 miles
   - Ⓒ 150 miles
   - Ⓓ 200 miles

5. Which highway would you use to get from Watertown, South Dakota, to Sioux City, Iowa?
   - Ⓐ Interstate Highway 29
   - Ⓑ Interstate Highway 90
   - Ⓒ United States Highway 18
   - Ⓓ United States Highway 212

6. What would be the best route to get from Rapid City to Brookings?
   - Ⓐ Interstate Highway 90
   - Ⓑ State Highway 79 and United States Highway 18
   - Ⓒ Interstate Highway 90 and United States Highway 14
   - Ⓓ State Highway 20

Name _____ Date _____

UNIT 3 PRE-ASSESSMENT

Directions: Use the map below and Map J. Darken the circle by the correct answer to each question.

1. Which of these states is in the Northeast Region of the United States?
   - Ⓐ North Dakota
   - Ⓑ Nevada
   - Ⓒ New Jersey
   - Ⓓ Arkansas

2. Which of these states is in the Southeast Region of the United States?
   - Ⓐ Arizona
   - Ⓑ Georgia
   - Ⓒ Idaho
   - Ⓓ Michigan

3. Which of these states is in the North Central Region of the United States?
   - Ⓐ Texas
   - Ⓑ Minnesota
   - Ⓒ North Carolina
   - Ⓓ California

4. Which of these states is in the Southwest Region of the United States?
   - Ⓐ New Mexico
   - Ⓑ Oregon
   - Ⓒ Wisconsin
   - Ⓓ West Virginia

5. Which of these states is in the Pacific Region of the United States?
   - Ⓐ Iowa
   - Ⓑ Maine
   - Ⓒ Florida
   - Ⓓ California

6. Utah is in the Rocky Mountain Region of the United States. Which of these states would also be in the Rocky Mountain Region?
   - Ⓐ Ohio
   - Ⓑ Delaware
   - Ⓒ Colorado
   - Ⓓ Alabama

Name _____ Date _____

UNIT 3 POST-ASSESSMENT

Directions: Use the map to answer the questions. Darken the circle by the correct answer to each question.

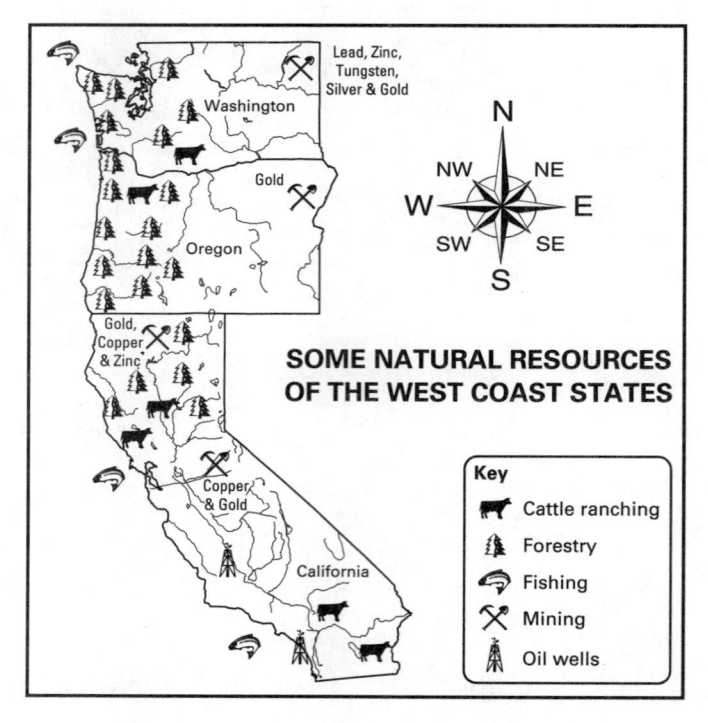

SOME NATURAL RESOURCES OF THE WEST COAST STATES

Key:
- Cattle ranching
- Forestry
- Fishing
- Mining
- Oil wells

1. Which states are shown on this map?
   - Ⓐ Wisconsin, Oregon, and California
   - Ⓑ Tungsten, Silver, and Gold
   - Ⓒ Gold, Copper, and Zinc
   - Ⓓ California, Washington, and Oregon

2. Where in the United States are these states located?
   - Ⓐ East Coast
   - Ⓑ West Coast
   - Ⓒ Gulf Coast
   - Ⓓ Gold Coast

3. What mineral is mined in eastern Oregon?
   - Ⓐ lead
   - Ⓑ zinc
   - Ⓒ copper
   - Ⓓ gold

4. What part of California uses land for forestry?
   - Ⓐ north
   - Ⓑ south
   - Ⓒ southwest
   - Ⓓ southeast

5. Which state has oil as one of its natural resources?
   - Ⓐ California
   - Ⓑ Oregon
   - Ⓒ Washington
   - Ⓓ none of these

6. What are two important resources found along the coast of the three states?
   - Ⓐ mining and oil
   - Ⓑ fishing and forestry
   - Ⓒ cattle and mining
   - Ⓓ none of these

Name _____ Date _____

UNIT 4 PRE-ASSESSMENT

Directions: Use the map to answer the questions. Darken the circle by the correct answer to each question.

[Map of THE UNITED STATES showing all 50 states labeled, with Canada to the north and Mexico to the south, and a compass rose showing N, S, E, W.]

1. Which state is directly west of Iowa?
   - Ⓐ Illinois
   - Ⓑ Nebraska
   - Ⓒ Missouri
   - Ⓓ Minnesota

2. Which state is directly north of Oregon?
   - Ⓐ California
   - Ⓑ Idaho
   - Ⓒ Washington
   - Ⓓ Montana

3. Which state is directly north of South Dakota?
   - Ⓐ Montana
   - Ⓑ Wyoming
   - Ⓒ North Carolina
   - Ⓓ North Dakota

4. Which of these states is east of Ohio?
   - Ⓐ Indiana
   - Ⓑ Missouri
   - Ⓒ Wisconsin
   - Ⓓ Maryland

5. Which of these states is an island?
   - Ⓐ Hawaii
   - Ⓑ Florida
   - Ⓒ Maine
   - Ⓓ Alaska

6. Which direction would you travel to get from Wisconsin to New York?
   - Ⓐ north
   - Ⓑ south
   - Ⓒ east
   - Ⓓ west

Name _____ Date _____

UNIT 4 POST-ASSESSMENT

Directions: Use the map to answer the questions. Darken the circle by the correct answer to each question.

[Map of THE UNITED STATES showing all 50 states labeled, with Canada to the north and Mexico to the south. A compass rose shows N, S, E, W directions.]

1. How many states are in the United States?
   Ⓐ 40
   Ⓑ 48
   Ⓒ 50
   Ⓓ 60

2. Which three states touch the southwest corner of Colorado?
   Ⓐ Kansas, Oklahoma, and Texas
   Ⓑ Utah, Arizona, and New Mexico
   Ⓒ Utah, Wyoming, and Idaho
   Ⓓ Wyoming, Nebraska, and Kansas

3. Which of these states touches one of the Great Lakes?
   Ⓐ Ohio
   Ⓑ Iowa
   Ⓒ Kentucky
   Ⓓ Virginia

4. Which of these states shares a border with Mexico?
   Ⓐ Montana
   Ⓑ Louisiana
   Ⓒ Hawaii
   Ⓓ Texas

5. Which direction would you travel to get from Illinois to South Dakota?
   Ⓐ northeast
   Ⓑ northwest
   Ⓒ southwest
   Ⓓ southeast

6. Which of these states is a peninsula?
   Ⓐ Hawaii
   Ⓑ Missouri
   Ⓒ Florida
   Ⓓ Pennsylvania

www.svschoolsupply.com
© Steck-Vaughn Company

Assessments
Weekly Geography Practice 3, SV 3432-0

Name _____ Date _____

UNIT 5 PRE-ASSESSMENT

Directions: Use the map to answer the questions. Darken the circle by the correct answer to each question.

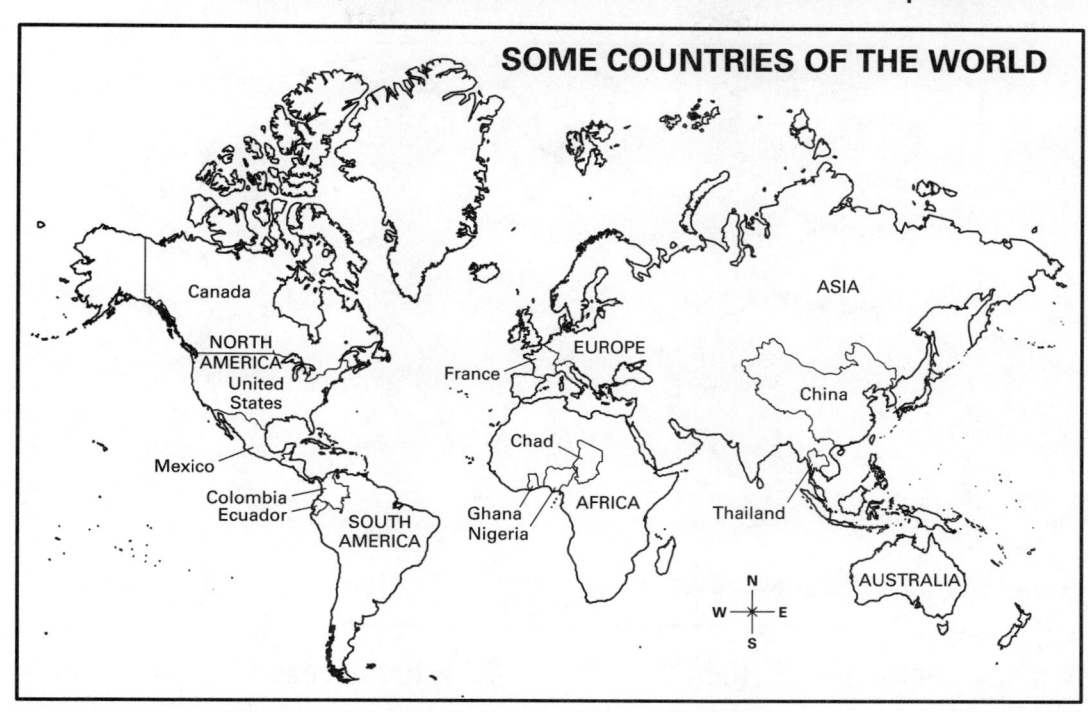

SOME COUNTRIES OF THE WORLD

1. On what continent is the United States located?
   - Ⓐ Europe
   - Ⓑ North America
   - Ⓒ South America
   - Ⓓ Africa

2. On what continent is Nigeria located?
   - Ⓐ Africa
   - Ⓑ Asia
   - Ⓒ Europe
   - Ⓓ Australia

3. On what continent is China located?
   - Ⓐ Europe
   - Ⓑ Africa
   - Ⓒ Asia
   - Ⓓ Australia

4. On what continent is Ecuador located?
   - Ⓐ Europe
   - Ⓑ Asia
   - Ⓒ South America
   - Ⓓ Australia

5. To travel directly from North America to Europe, which direction must you go?
   - Ⓐ north
   - Ⓑ south
   - Ⓒ east
   - Ⓓ west

6. To travel directly from Europe to South America, which direction must you go?
   - Ⓐ northwest
   - Ⓑ southeast
   - Ⓒ northeast
   - Ⓓ southwest

Name _____ Date _____

UNIT 5 POST-ASSESSMENT

Directions ➤ Use the map to answer the questions. Darken the circle by the correct answer to each question.

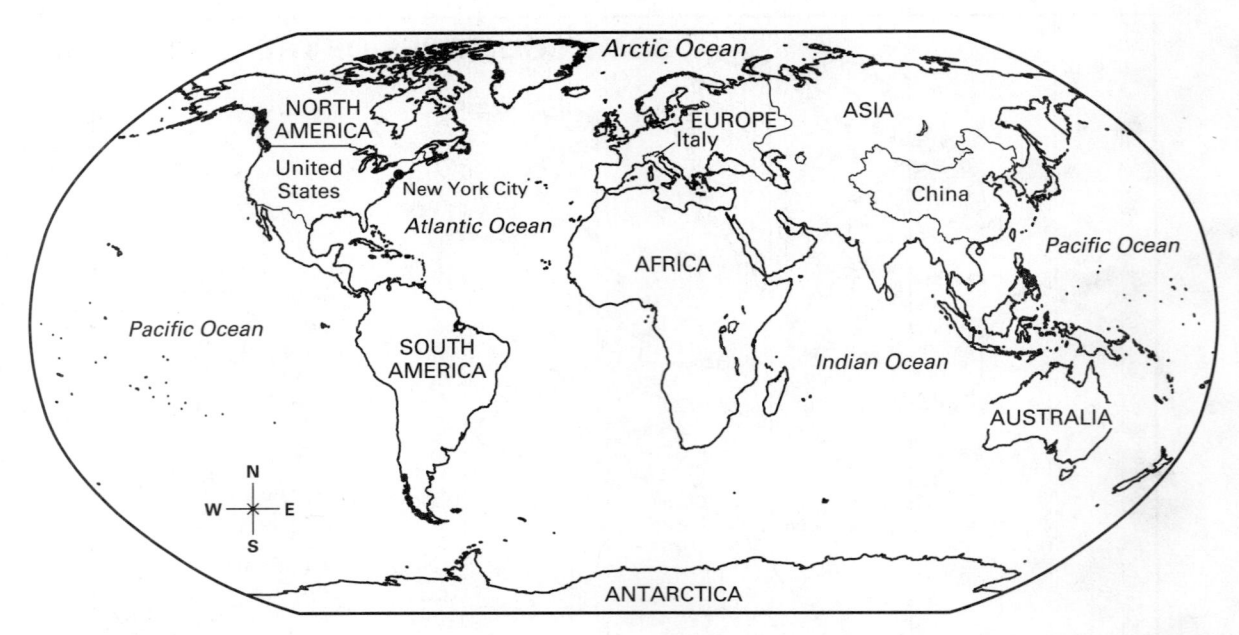

1. How many continents are there?
   - Ⓐ five
   - Ⓑ six
   - Ⓒ seven
   - Ⓓ ten

2. To travel directly from the United States to Europe, which ocean must you cross?
   - Ⓐ Arctic Ocean
   - Ⓑ Atlantic Ocean
   - Ⓒ Pacific Ocean
   - Ⓓ Indian Ocean

3. To travel directly from Africa to Australia, which ocean must you cross?
   - Ⓐ Arctic Ocean
   - Ⓑ Atlantic Ocean
   - Ⓒ Pacific Ocean
   - Ⓓ Indian Ocean

4. Which ocean is the most northern?
   - Ⓐ Arctic Ocean
   - Ⓑ Atlantic Ocean
   - Ⓒ Pacific Ocean
   - Ⓓ Indian Ocean

5. Which ocean touches both South America and Asia?
   - Ⓐ Arctic Ocean
   - Ⓑ Atlantic Ocean
   - Ⓒ Pacific Ocean
   - Ⓓ Indian Ocean

6. To travel directly from Europe to Australia, which direction must you go?
   - Ⓐ northwest
   - Ⓑ southeast
   - Ⓒ northeast
   - Ⓓ southwest

Name: Caleb   Date: Oct. 2003

Monday

1. What is the name of your community?

   "Sherman Valley"

2. What is your full address and telephone number?

   9110 Sherman Valley Rd SW
   Olympia, W.A. 98510
   (360) 556-5410

Tuesday

1. Do you live in or near a town, a small city, or a big city?

   We live in Capital forest.

2. Do you live near an ocean, a lake or river, a forest, a desert, or mountains?

   forest and creek

Wednesday

1. How do you get to school each day?

   I walk in the trailer to Homeschool.

2. About how far do you live from your school?

   100 feet away.

Thursday

1. What fun things are there to do in your community?

   fish, hunt, ride motorcycles, hike in woods, collect animal insects

2. Name three things you think make a community a good place to live.

   house, food, toys

Friday

1. Why do you think people chose this place to start your community?

   Because it is quiet

2. Do you know of a pollution problem in your community? What can people do to stop pollution?

   no we live in the forest —

Week Number 1 2 3 4 5 6 7 8 9 10 11 12 13 14 15 16 17 18 19 20 21 22 23 24 25 26 27 28 29 30 31 32 33 34 35 36

Focus: The purpose of this week's questions is to ready the students for some of the concepts associated with geography and maps, such as location, distance, movement, and human interaction with the environment.

Monday

1. Answer: Answers will vary.
If you know the origin of the community's name, share it with the students.

2. Answer: Answers will vary.
Review the various parts of an address, such as street name and number, city, state, and ZIP code. See if the students know the area code for their telephone number.

Tuesday

1. Answer: Answers will vary.
Point out to the students that communities come in many different sizes.

2. Answer: Answers will vary.
Point out to the students that communities are founded in many different places, with many differing kinds of surroundings or landforms.

Wednesday

1. Answer: Answers will vary.
Review with the students the many different modes of transportation. Perhaps take a poll to see which mode is most used.

2. Answer: Answers will vary.
Point out to the students that ease of movement is often related to the distance to a place.

Thursday

1. Answer: Answers will vary.
Point out to the students that recreational opportunities are an important part of a community.

2. Answer: Answers will vary.
Write all the student responses on the board. Responses might include safety, job opportunities, good schools, good water, recreational opportunities, etc. Take a vote to see which three the students believe are the most important.

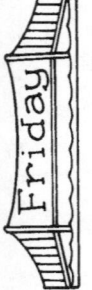

Friday

1. Answer: Answers will vary.
Point out to the students that many communities are founded based on a good location. Introduce the concept of the history of the community. Challenge the students to find information about Native American groups that might have lived in the area before the community was settled.

2. Answer: Answers will vary.
Clean air and clean water are two important parts of a good quality of life in a community. Point out to the students that people must work together to prevent pollution.

Week Number 1 2 3 4 5 6 7 8 9 10 11 12 13 14 15 16 17 18 19 20 21 22 23 24 25 26 27 28 29 30 31 32 33 34 35 36

Name _____ Date _____

Directions → Use **Map A** to help you answer the questions.

Monday

1. What can a map show you?
_____
_____
_____

2. What part of a map tells you direction? What are the four cardinal directions?
_____
_____
_____

Tuesday

1. What are the pictures or colors on a map that stand for real things?
_____
_____
_____

2. What is the list of symbols on a map called?
_____
_____
_____

Wednesday

1. What part of a map helps you know how far places really are from each other?
_____
_____

2. What are some ways that people can move from one place to another?
_____
_____
_____

Thursday

1. What kinds of different things are shown on this map?
_____
_____
_____

2. Why is it good to know about other places?
_____
_____
_____

Friday

1. What might people do for fun in this town?
_____
_____
_____

2. Why is water an important resource for a community?
_____
_____
_____

Week Number 1 2 3 4 5 6 7 8 9 10 11 12 13 14 15 16 17 18 19 20 21 22 23 24 25 26 27 28 29 30 31 32 33 34 35 36

www.svschoolsupply.com
© Steck-Vaughn Company

Unit 1: Community
Weekly Geography Practice 3, SV 3432-0

Focus: The purpose of this week's questions is to acquaint students with the general features of a map and to make them think beyond the simple physical characteristics of a map.

Directions: Distribute **Map A** to assist students in answering the questions.

Monday

1. Answer: where things are located.
Point out that there are many kinds of maps: political; physical; population; climate and weather; route; resource; cultural. All maps have as their purpose the attempt to help people to understand an area better.

2. Answer: compass rose or direction marker; north, east, south, west.
Point out that the compass rose allows the map user to know which direction is which. Direction is important in getting from one place to another. You might also note the four intermediate directions: northeast, southeast, southwest, northwest.

Tuesday

1. Answer: symbols.
The symbols are a graphical way of informing the map user where various features are located.

2. Answer: map key or map legend.
The map key is important in allowing the map user to know what all the symbols stand for. Without the map key, the user might not know what the symbols are meant to represent. Then, the map would have little value.

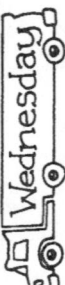

Wednesday

1. Answer: distance scale.
Point out that maps are not full-size. They are scaled. The distance scale allows the user to estimate real distances. Usually the distances are given in miles or kilometers. On **Map A**, one inch on the map equals three miles in real distance.

2. Answer: Answers will vary, but should include the common modes of transportation—by road, by rail, by water, by air, by walking, by using animals.
Point out that people have developed many ways to move from one place to another. Indicate that the physical features of an area may determine the best mode of transportation.

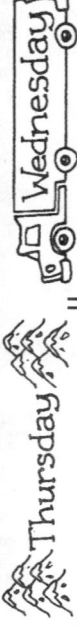

Thursday

1. Answer: Map A shows various features of Bend River, such as roads, bridges, railroad tracks, railroad station, a river, and a larger body of water.
Point out that the map helps the user to understand how the various features are situated in relation to one another.

2. Answer: Answers will vary, but should suggest that the knowledge of other places gives one a better understanding of the world in which one lives.
Point out that knowing about other places can help the students to have a better understanding of the place in which they live, simply through comparison and contrast.

Friday

1. Answer: The presence of a river and a larger body of water suggests that water activities might be a type of entertainment in this community.
Point out that recreational fun is often determined by the physical features of an area.

2. Answer: Answers will vary, but should suggest that water is necessary for life.
People in a community use water to drink, to cook, to clean, to nurture plants and animals, for transportation, for industry, and for recreation.

Week Number 1 2 3 4 5 6 7 8 9 10 11 12 13 14 15 16 17 18 19 20 21 22 23 24 25 26 27 28 29 30 31 32 33 34 35 36

Name _____ Date _____

Directions: Use **Map A** to help you answer the questions.

Monday

1. What is the name of this town?

2. What is the symbol for a hospital?

Tuesday

1. Is the factory east or west of the river?

2. Is the hospital north or south of the school?

Wednesday

1. How many bridges are in this town?

2. People in this town can travel on the streets or on the railroad. How else can they travel?

Thursday

1. What is the distance between the two bridges on the river?

2. About how far is the school from the railroad station?

Friday

1. Where could people work in this town?

2. What can people do to take care of this community?

Week Number 1 2 3 4 5 6 7 8 9 10 11 12 13 14 15 16 17 18 19 20 21 22 23 24 25 26 27 28 29 30 31 32 33 34 35 36

www.svschoolsupply.com
© Steck-Vaughn Company

Unit 1: Community
Weekly Geography Practice 3, SV 3432-0

Focus: The purpose of this week's questions is to acquaint students with the specific features of a map, especially direction and distance.

Directions: Distribute **Map A** to assist students in answering the questions.

Monday

1. **Answer:** Bend River.
   Point out that most maps have titles that tell what the map is about.

2. **Answer:** a cross.
   Point out that students should check the map key to see what the symbols on a map represent.

Tuesday

1. **Answer:** The factory is east of the river.
   Point out that easy movement can be restricted by physical features. Here, people must cross a river bridge to get to the factory.

2. **Answer:** The hospital is north of the school.
   Point out that directions are important to movement.

Wednesday

1. **Answer:** three bridges.
   Point out that the bridges are important to allow easier movement.

2. **Answer:** They can travel by water or overland by walking or by horseback, for example.
   Some students might suggest travel by air, but point out that the map of this community does not show an airport.

Thursday

1. **Answer:** five to six miles.
   Point out that the river restricts easy movement. People can usually cross the river only at the bridges, so knowing the placement of the bridges and the distance between them allows one to plan travel better.

2. **Answer:** four to five miles.
   Point out that knowing the distance between places allows one to calculate travel time better.

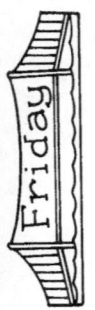

Friday

1. **Answer:** Three obvious answers are at the school, at the railroad station, and at the factory.
   Point out that other people might work as mail carriers or delivery people, as house painters or lawn mowers. There are many less obvious jobs that the students might suggest.

2. **Answer:** Answers will vary, but might suggest that the people can keep their community clean and safe, and that they can work to prevent air and water pollution.

Week Number 1 2 3 4 5 6 7 8 9 10 11 12 13 14 15 16 17 18 19 20 21 22 23 24 25 26 27 28 29 30 31 32 33 34 35 36

www.svschoolsupply.com
© Steck-Vaughn Company

Unit 1: Community
Weekly Geography Practice 3, SV 3432-0

Name _____ Date _____

Directions → Use **Map B** to help you answer the questions.

Monday

1. On the map of Oak Bluff, the tree is the symbol for what place?
_____
_____

2. This map uses a grid. It has eight sections. Each section is identified by a letter and a number. What kind of place is located in grid section B-2?
_____
_____

Tuesday

1. On what avenue is the library located?
_____
_____

2. Is the school on Second Avenue north or south of the library?
_____
_____

Wednesday

1. To travel from Oak Street to Elm Street, would you go east or west?
_____
_____

2. Which route would you use to get from the library to the school in grid section D-1?
_____
_____

Thursday

1. On which street are all of the businesses located?
_____
_____

2. How many schools are in Oak Bluff?
_____
_____

Friday

1. Which grid section contains the most park area?
_____
_____

2. Why are parks good for a community?
_____
_____

Week Number 1 2 3 4 5 6 7 8 9 10 11 12 13 14 15 16 17 18 19 20 21 22 23 24 25 26 27 28 29 30 31 32 33 34 35 36

Focus: The purpose of this week's questions is to acquaint students with the specific features of a map, especially direction, highways, and grid sections.

Directions: Distribute **Map B** to assist students in answering the questions.

Monday

1. Answer: park.
Review the use of symbols in the map key to represent places on the map.

2. Answer: school.
Demonstrate to the students that this map is divided into a grid of eight sections. Point out that the grid sections are identified by a letter and number combination, called coordinates, such as A-2. Many map indexes name place locations by using grid coordinates.

Tuesday

1. Answer: First Avenue.

2. Answer: south.
You might want to review the cardinal directions and point out the compass rose to the students.

Wednesday

1. Answer: east.

2. Answer: Answers may vary. One route is First Avenue to Oak Street to the school. Another route is First Avenue to Main Street to Fourth Avenue to the school.
Accept all possible routes. Point out, though, that the most direct route is often the fastest.

Thursday

1. Answer: Third Avenue.
Some students might say Second Avenue. Point out that buildings usually have addresses on the street they face. Also note that businesses are often grouped together in the "business section" of a community.

2. Answer: two.

Friday

1. Answer: A-1.
Sections A-2 and D-2 also have parkland. Suggest that because A-1 has more tree symbols, it has more parkland.

2. Answer: Answers will vary, but should suggest that parks add greenery and recreational opportunities to a community, as well as promote the quality of life.

Week Number 1 2 3 4 5 6 7 8 9 10 11 12 13 14 15 16 17 18 19 20 21 22 23 24 25 26 27 28 29 30 31 **32** 33 34 35 36

www.svschoolsupply.com
© Steck-Vaughn Company

Unit 1: Community
Weekly Geography Practice 3, SV 3432-0

Name _____ Date _____

Directions: Use **Map B** to help you answer the questions.

Monday

1. In which grid section is the library located?

2. In which two grid sections are all the houses located?

Tuesday

1. Which direction is the library from the school in the southwest corner of Oak Bluff?

2. The corner where Elm Street meets Third Avenue is which direction from the houses on Second Avenue?

Wednesday

1. A school is located southeast of the corner where which two streets meet?

2. To travel from the library to the school in D-1, which direction must you go?

Thursday

1. Are there more businesses in grid section C-1 or C-2?

2. How do grid sections make a map easier to use?

Friday

1. Why is the library an important place in a community?

2. What kinds of books do you like to read? Who is your favorite author?

Week Number 1 2 3 4 5 6 7 8 9 10 11 12 13 14 15 16 17 18 19 20 21 22 23 24 25 26 27 28 29 30 31 32 33 34 35 36

Focus: The purpose of this week's questions is to acquaint students with the specific features of a map, especially direction, street names, and grid sections.

Directions: Distribute **Map B** to assist students in answering the questions.

Monday

1. **Answer:** A-2.
   Review the use of coordinates.

2. **Answer:** B-1 and D-2.

Tuesday

1. **Answer:** northeast.
   Point out the intermediate directions on the compass rose. Also point out to the students that they must sometimes be aware of several directional descriptions to locate a place.

2. **Answer:** southeast.
   Point out to the students that directions are often given that refer to the intersection of streets or roads.

Wednesday

1. **Answer:** Main Street and First Avenue.
   Some students may give the answer Oak Street and Third (or Second) Avenue, which is also correct because there are two schools.

2. **Answer:** southwest.

Thursday

1. **Answer:** C-1.
   Point out that there are four business symbols in C-1 and only three business symbols in C-2.

2. **Answer:** Grid sections break the whole map into smaller parts. Places are easier to find in the smaller parts.

Friday

1. **Answer:** The library provides all kinds of resources so the people of the community can learn more or be entertained.

2. **Answer:** Answers will vary.
   Take a survey to find which authors are the most popular among the students.

Name _____ Date _____

Directions Use **Map C** to help you answer the questions.

Monday

1. In which grid section is City Hall located?
   _____
   _____

2. In which grid section is Acme Supermarket located?
   _____
   _____

Tuesday

1. What can be found at the intersection of Main Street and Oak Street?
   _____
   _____

2. The police station is located at the intersection of what two streets?
   _____
   _____

Wednesday

1. To travel from the high school to the elementary school, which direction must you go?
   _____
   _____

2. To travel from the fire station to the police station, which direction must you go?
   _____
   _____

Thursday

1. What is the name of the city shown on this map?
   _____
   _____

2. What are the names of the two lakes shown on the map?
   _____
   _____

Friday

1. Why do you think Horseshoe Lake was given that name?
   _____
   _____

2. How can the people of this city use their two lakes?
   _____
   _____

Week Number 1 2 3 4 5 6 7 8 9 10 11 12 13 14 15 16 17 18 19 20 21 22 23 24 25 26 27 28 29 30 31 32 33 34 35 36

www.svschoolsupply.com
© Steck-Vaughn Company

Unit 1: Community
Weekly Geography Practice 3, SV 3432-0

Focus: The purpose of this week's questions is to continue to acquaint the students with the specific features of a map, especially direction, street names, and grid sections.

Directions: Distribute **Map C** to assist students in answering the questions.

Monday

1. Answer: A-1.
Point out to the students that this map has a grid made up of 12 sections.

2. Answer: B-2.
Point out the location to the students.

Tuesday

1. Answer: the post office.
Show the students how to locate the intersection. If necessary, define an intersection as the place where two or more streets meet or cross one another.

2. Answer: Main Street and Olive Street.

Wednesday

1. Answer: south.
This would be a good time to stress the importance of traffic safety, especially for pedestrians. Ask the students if they know how to use crosswalks at intersections.

2. Answer: northwest.
You can continue the discussion of traffic safety by reviewing bike and bus safety.

Thursday

1. Answer: Smithville.
Review the importance of titles on maps.

2. Answer: Horseshoe Lake and Mirror Lake.
Point out the two lakes to the students.

Friday

1. Answer: It is shaped like a horseshoe.

2. Answer: Answers will vary, but may include swimming, wading, boating, fishing, etc.

Week Number 1 2 3 4 5 6 7 8 9 10 11 12 13 14 15 16 17 18 19 20 21 22 23 24 25 26 27 28 29 30 31 32 33 34 35 36

www.svschoolsupply.com
© Steck-Vaughn Company

36

Unit 1: Community
Weekly Geography Practice 3, SV 3432-0

Name _____ Date _____

Directions → Use **Map C** to help you answer the questions.

Monday

1. In which grid section is the hospital located?
_____
_____

2. In which two grid sections is Horseshoe Lake located?
_____
_____

Tuesday

1. What can be found at the intersection of Oak Street and Capital Avenue?
_____
_____

2. The fire station is located at the intersection of what two streets?
_____
_____

Wednesday

1. What route would you use to travel from the mini-mall to the hospital?
_____
_____

2. What route would you use to travel from the police station to Mirror Lake?
_____
_____

Thursday

1. What are the names of the two schools in this city?
_____
_____

2. In what part of the city are the two schools located?
_____
_____

Friday

1. This city has two parks. What are their names?
_____
_____

2. What things do you like to do in the park?
_____
_____

Week Number 1 2 3 4 5 6 7 8 9 10 11 12 13 14 15 16 17 18 19 20 21 22 23 24 25 26 27 28 29 30 31 32 33 34 35 36

Focus: The purpose of this week's questions is to continue to acquaint the students with the specific features of a map, especially direction, street names, and grid sections.

Directions: Distribute **Map C** to assist students in answering the questions.

Monday

1. Answer: C-3.

2. Answer: A-2 and B-2.

Tuesday

1. Answer: the mini-mall.
Some students might answer Hubbard Park, but the mini-mall is closer to the intersection.

2. Answer: Capital Avenue and Main Street.
Point out the location to the students.

Wednesday

1. Answer: Capital Avenue to Main Street to the hospital.

2. Answer: Main Street to Oak Street to Capital Avenue to the park and lake.
An acceptable alternate route would be Main Street to Capital Avenue to the park and lake.

Thursday

1. Answer: Jefferson High School and Washington Elementary School.

2. Answer: east or northeast part of the city.

Friday

1. Answer: Veteran's Park and Hubbard Park.

2. Answer: Answers will vary.
Take a survey to see what the students' favorite activities are.

Name _____ Date _____

Directions → Use **Map D** to help you answer the questions.

Monday

1. What is the name of the city that is shaded on this map?
   _____
   _____

2. What city is located on State Highway 4?
   _____
   _____

Tuesday

1. What is on the east border of Willow?
   _____
   _____

2. What two suburbs are west of Willow?
   _____
   _____
   _____

Wednesday

1. About how far is State Highway 4 from the river bridge?
   _____
   _____

2. In which direction would you travel from Willow to the farmland?
   _____
   _____

Thursday

1. What separates Hogan from Dixon?
   _____
   _____

2. What separates Willow from its two suburbs?
   _____
   _____
   _____

Friday

1. How can people cross the river shown on the map?
   _____
   _____

2. Why might people choose to live in a suburb instead of a city?
   _____
   _____
   _____

Focus: The purpose of this week's questions is to acquaint the students with the concept of a community's surrounding areas, such as suburbs.

Directions: Distribute **Map D** to assist students in answering the questions.

Monday

1. Answer: Willow.

2. Answer: Keyville.
Point out this city to the students.

Tuesday

1. Answer: Deep River.
Point out the river to the students.

2. Answer: Hogan and Dixon.

Wednesday

1. Answer: about 25 miles.
Review the use of the distance scale.

2. Answer: west.

Thursday

1. Answer: U.S. Highway 24.

2. Answer: U.S. Highway 51.

Friday

1. Answer: The best answer would be "using the bridge."

2. Answer: Answers will vary, but may include ideas such as less traffic, fewer people, a better quality of life, etc.

Name _____ Date _____

Directions → Use **Map E** to help you answer the questions.

Monday

1. What is the name of the river shown on the map?

2. What is the only interstate highway shown on the map?

Tuesday

1. In what general direction does State Highway 50 run?

2. Which city is located twelve miles east and nine miles northeast of Schnell?

Wednesday

1. What is the shortest distance from Collier to Kaimen?

2. What would be the shortest route from Bullock to Schnell?

Thursday

1. Which highways are shown crossing the river?

2. What is the largest city in Jones County?

Friday

1. What is the population of O'Reilly?

2. Which city might have the most jobs available?

Week Number 1 2 3 4 5 6 7 8 9 10 11 12 13 14 15 16 17 18 19 20 21 22 23 24 25 26 27 28 29 30 31 32 33 34 35 36

Focus: The purpose of this week's questions is to acquaint students with the specific features of a map, especially direction, rivers, highways, and city sizes.

Directions: Distribute **Map E** to assist students in answering the questions.

Monday

1. Answer: Souby River.
Point out to the students that rivers are usually named on the map and differ in appearance from highways.

2. Answer: Interstate 3.
Point out to the students that there are several classifications of highways, including interstate highways, U.S. highways, and state highways.

Tuesday

1. Answer: southeast to northwest (or northwest to southeast).
Point out to the students that highways often continue in a general direction along their entire length, even though they may twist and curve at times.

2. Answer: Lesford.
Point out to the students that distance on this map is indicated by a number situated between arrows, instead of by a distance scale. Indicate the difference between these mileage indicators and the highway numbers, which are enclosed in some kind of outline.

Wednesday

1. Answer: 16 miles.
Point out that there are at least two possible routes, but one is shorter than the other.

2. Answer: southward along Interstate 3 to Russell, then west on State Highway 41 to Schnell, a distance of 44 miles.
Point out that other routes are possible, but the one above is the shortest.

Thursday

1. Answer: as shown on the map, Interstate 3, U.S. Highway 13, and State Highway 41 cross the river.
Some of the other highways may cross the river, but they are not shown on the map doing so.

2. Answer: Russell.
Point out to the students the symbols that represent city size in population.

Friday

1. Answer: less than 1,000 people.
Point out to the students that people like to live in a place for different reasons. As a result, some cities have larger populations than other cities.

2. Answer: Russell.
Suggest to the students that because Russell has the largest population, it very likely also has the largest number of jobs available.

Week Number 1 2 3 4 5 6 7 8 9 10 11 12 13 14 15 16 17 18 19 20 21 22 23 24 25 26 27 28 29 30 31 32 33 34 35 36

Name _____ Date _____

Directions: Use **Map G** to help you answer the questions.

Monday

1. The United States is made up of 50 parts. What are these parts called? What is the name of your part?

_____
_____

2. States in the United States are made of many parts. What are these parts called? What is the name of your part?

_____
_____

Tuesday

1. What other states touch the border of your state?

_____
_____

2. Does your state touch the border of another country? If so, which country?

_____
_____

Wednesday

1. Is your state near an ocean, a desert, or mountains? How might these places help or hurt transportation?

_____
_____

2. Are there any large rivers in your state? How are they used?

_____
_____

Thursday

1. Look at the map. Is your state a big state, a medium-size state, or a small state?

_____
_____

2. Do you have a favorite sports team in your state? What is the team?

_____
_____

Friday

1. What are some of the important jobs in your state?

_____
_____

2. Do you know of any pollution problems in your state? How might they be solved?

_____
_____

Week Number 1 2 3 4 5 6 7 8 9 10 11 12 13 14 15 16 17 18 19 20 21 22 23 24 25 26 27 28 29 30 31 32 33 34 35 36

www.svschoolsupply.com

© Steck-Vaughn Company

Unit 2: State
Weekly Geography Practice 3, SV 3432-0

Focus: The purpose of this week's questions is to introduce the students to concepts associated with the larger political division of the state.

Directions: Distribute **Map G** to assist students in answering the questions.

Monday

1. **Answer:** states; Answers will vary.
   If you know the origin of the state's name, share it with the students. If you know the motto, teach it to the students.

2. **Answer:** counties; Answers will vary.
   If you know the origin of the county's name, share it with the students. Note that not all states use the county as a basic political division. Louisiana, for example, has parishes instead of counties.

Tuesday

1. **Answer:** Answers will vary.
   Point out that a state often has important links to a neighboring state, such as in economy or through regional similarities.

2. **Answer:** Answers will vary.
   Point out that states that share a border with Mexico or Canada often have important links to that neighboring country, such as in economy or through immigration.

Wednesday

1. **Answer:** Answers will vary.
   Point out that oceans allow better transportation, and that mountains and deserts usually serve as obstacles to transportation.

2. **Answer:** Answers will vary.
   Rivers are usually used for transporting people and goods, and for recreational purposes, such as swimming, boating, and fishing. Some rivers are also used to generate electrical power or for agricultural irrigation.

Thursday

1. **Answer:** Answers will vary.
   Point out to the students that states come in many different sizes.

2. **Answer:** Answers will vary.
   Point out that sports teams often have mascots that reflect the local or state history. Sports teams also may serve to promote good-natured regional rivalries.

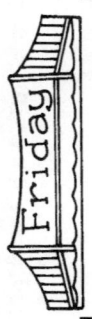

Friday

1. **Answer:** Answers will vary.
   Point out that the healthy economy of a state depends on its people having good jobs.

2. **Answer:** Answers will vary.
   Most states have some sort of pollution problem. Survey the students for innovative solutions to pollution problems.

Week Number 1 2 3 4 5 6 7 8 9 10 11 12 13 14 15 16 17 18 19 20 21 22 23 24 25 26 27 28 29 30 31 32 33 34 35 36

Name _____ Date _____

Directions → Use **Map G** to help you answer the questions.

Monday

1. What state borders North Dakota to the east?

2. What state borders Alabama to the north?

Tuesday

1. Which state borders only one other state?

2. Which state is farther west: North Carolina or South Carolina?

Wednesday

1. To get directly from Illinois to Kansas, which state would a traveler cross?

2. In which direction would you travel going from Montana to Colorado?

Thursday

1. Which state is divided into two large parts, with one part not attached to the other part?

2. Which two states have four sides, all of which are straight?

Friday

1. Where are some good places for recreation, such as camping or swimming, in your state?

2. Name three things that make your state a good place to live.

Focus: The purpose of this week's questions is to acquaint the students with various states in the United States and to reinforce the concept of direction.

Directions: Distribute **Map G** to assist students in answering the questions.

Monday

1. **Answer:** Minnesota.
You might ask the students to identify states bordering North Dakota in the other cardinal directions. Canada borders North Dakota on the north.

2. **Answer:** Tennessee.
You might ask the students to identify states bordering Alabama in the other cardinal directions. Point out that Alabama has a partial border on the Gulf of Mexico.

Tuesday

1. **Answer:** Maine.
Its position in the far northeast corner of the United States limits it to a border only with New Hampshire. Maine, though, does have a border on the Atlantic Ocean, as well as a border with Canada.

2. **Answer:** North Carolina.
Point out how North Carolina extends farther west than South Carolina. Ask the students what borders these two states on the east (Atlantic Ocean).

Wednesday

1. **Answer:** Missouri.
Point out that a traveler would also have to cross the Mississippi River to reach Kansas.

2. **Answer:** south.
Review the cardinal directions. Ask the students which direction they would need to travel to get from Colorado to Montana (north).

Thursday

1. **Answer:** Michigan.
The state is divided by Lake Michigan. Point out that the lower half of the state looks like a glove.

2. **Answer:** Colorado and Wyoming.
Point out that the states have many different shapes. Several states, such as Texas, Oklahoma, Idaho, and Florida, have a shape called a panhandle. See if the students can find these shapes on their map.

Friday

1. **Answer:** Answers will vary.
Ask the students if they like to take vacations, and why. What is their favorite vacation spot?

2. **Answer:** Answers will vary.
Make a list of all the student responses. Take a vote to determine the three most important things.

Week Number 1 2 3 4 5 6 7 8 9 10 11 12 13 14 15 16 17 18 19 20 21 22 23 24 25 26 27 28 29 30 31 32 33 34 35 36

www.svschoolsupply.com
© Steck-Vaughn Company

Unit 2: State
Weekly Geography Practice 3, SV 3432-0

Name _____ Date _____

Directions: Use **Map G** to help you answer the questions.

Monday

1. If you were in Cheyenne, in which state's capital would you be?

2. Which of these cities is the capital of New York: Boston, Albany, or Trenton?

Tuesday

1. Which city is the capital of the state east of Mississippi and west of Georgia?

2. Which of these cities is not a state capital: Boise, Los Angeles, or Nashville?

Wednesday

1. Which direction would you travel to get from Charleston, West Virginia, to Raleigh, North Carolina?

2. Which direction would you travel to get from Austin, Texas, to Santa Fe, New Mexico?

Thursday

1. Of which two neighboring states are Lincoln and Topeka the capitals?

2. Chicago is the largest city in Illinois, but it is not the capital. Which city is the capital of Illinois?

Friday

1. Which capital city is located on an island?

2. Which capital city is named for a large, salty lake?

Week Number 1 2 3 4 5 6 7 8 9 10 11 12 13 14 15 16 17 18 19 20 21 22 23 24 25 26 27 28 29 30 31 32 33 34 35 36

Unit 2: State

Focus: The purpose of this week's questions is to acquaint the students with various state capitals.
Directions: Distribute **Map G** to assist students in answering the questions.

Monday

1. **Answer:** Wyoming.

2. **Answer:** Albany.

Tuesday

1. **Answer:** Montgomery, Alabama.
Review the importance of knowing directions on a map.

2. **Answer:** Los Angeles.
You might help students here by pointing out that Los Angeles is in California.

Wednesday

1. **Answer:** southeast.
Review the use of the compass rose, especially in regard to intermediate directions.

2. **Answer:** northwest.

Thursday

1. **Answer:** Nebraska (Lincoln) and Kansas (Topeka).

2. **Answer:** Springfield, Illinois.
Point out that the capital city in a state is often not the largest city.

Friday

1. **Answer:** Honolulu, Hawaii.
You might define the term *island* as land surrounded on all sides by water. The state of Hawaii is made up of many small islands. Honolulu is on the island called Oahu.

2. **Answer:** Salt Lake City, Utah.
Great Salt Lake is in the northwest part of Utah. Its waters are very salty. In fact, early explorers in the area thought the lake was part of the Pacific Ocean.

Name _____ Date _____

Directions → Use **Map G** to help you answer the questions.

Monday

1. What state borders Louisiana to the west?
   _____
   _____

2. What state borders Wyoming to the north?
   _____
   _____

Tuesday

1. Which state is farther north: Nevada or Nebraska?
   _____
   _____

2. Which state is farther west: Florida or Pennsylvania?
   _____
   _____

Wednesday

1. What is a path or road from one place to another called?
   _____
   _____

2. What is a place where two routes meet called?
   _____
   _____

Thursday

1. What state or body of water is directly east of Florida?
   _____
   _____

2. What state borders both Mexico and the Gulf of Mexico?
   _____
   _____

Friday

1. Why were many communities built at crossroads?
   _____
   _____

2. Why is it good to know things about other states?
   _____
   _____

Week Number 1 2 3 4 5 6 7 8 9 10 11 12 13 14 15 16 17 18 19 20 21 22 23 24 25 26 27 28 29 30 31 32 33 34 35 36

www.svschoolsupply.com

© Steck-Vaughn Company

Unit 2: State

Weekly Geography Practice 3, SV 3432-0

Focus: The purpose of this week's questions is to acquaint the students with various states in the United States, to reinforce the concept of direction, and to introduce the concept of routes.

Directions: Distribute **Map G** to assist students in answering the questions.

Monday

1. Answer: Texas.
You might ask the students to identify states bordering Louisiana in the other cardinal directions. Point out that Louisiana's southern border is the Gulf of Mexico.

2. Answer: Montana.
You might ask the students to identify states bordering Wyoming in the other cardinal directions.

Tuesday

1. Answer: Nebraska.
Point out that the northern border of Nebraska is slightly farther north than the northern border of Nevada.

2. Answer: Florida.
Point out that the panhandle of Florida is considerably farther west than the western border of Pennsylvania.

Wednesday

1. Answer: route.
The general term is *route*, though the students might give answers such as street, highway, trail, way, etc.

2. Answer: crossroads, crossing point, or intersection.
Perhaps you can use an example of a crossroads or intersection near the school.

Thursday

1. Answer: Atlantic Ocean.
Point out how Florida extends into the water. You might want to introduce the term *peninsula*. To the east of Florida is the Atlantic Ocean; to the west is the Gulf of Mexico.

2. Answer: Texas.
Review the different kinds of borders a state might have: another state, another country, a river or lake, an ocean or gulf, etc.

Friday

1. Answer: Crossing points were good stopping places for travelers, who sometimes decided to settle there.
Point out that many of the early routes and trails became the roads and highways of today.

2. Answer: Answers will vary.
Students might suggest that knowing about other states helps them to know more about their country and the different people that live in it.

Week Number 1 2 3 4 5 6 7 8 9 10 11 12 13 14 15 16 17 18 19 20 21 22 23 24 25 26 27 28 29 30 31 32 33 34 35 36

Name _____ Date _____

Directions ▸ Use **Map F** to help you answer the questions.

Monday

1. What two cities are directly west of Plainview?
_____
_____

2. Is Amarillo north or south of Lubbock?
_____
_____

Tuesday

1. What major east-west road passes through Amarillo?
_____
_____

2. Which city is farther west: Hereford or Plainview?
_____
_____

Wednesday

1. What river is shown on the map?
_____
_____

2. Name two routes to get from Muleshoe to Canyon.
_____
_____

Thursday

1. What is the name of this map?
_____
_____

2. What is the distance between Amarillo and Lubbock?
_____
_____

Friday

1. What state park is near Amarillo?
_____
_____

2. Why are good roads important for an area?
_____
_____

Week Number 1 2 3 4 5 6 7 8 9 10 11 12 13 14 15 16 17 18 19 20 21 22 23 24 25 26 27 28 29 30 31 32 33 34 35 36

Focus: The purpose of this week's questions is to acquaint the students with the use of a road map.
Directions: Distribute **Map F** to assist students in answering the questions.

Monday

1. Answer: Earth and Muleshoe.
Point out that many towns have unusual names. Ask the students if they can think of any.

2. Answer: north.
Review the use of the compass rose.

Tuesday

1. Answer: Interstate Highway 40.
Most interstate highways that end in an even number run east and west; most interstate highways that end in an odd number run north and south.

2. Answer: Hereford.
Hereford is more north and more west than Plainview.

Wednesday

1. Answer: Canadian River.
Point out that most roads use a bridge to cross rivers.

2. Answer: U.S. Highway 84 to U.S. Highway 60 to Canyon; U.S. Highway 70 to U.S. Highway 87 (or Interstate Highway 27) to Canyon.
Point out that other routes are possible, but these are the most direct.

Thursday

1. Answer: Road Map of West Texas.
Review the importance of a title on a map. Titles allow the map user to know what information the map contains.

2. Answer: about 120 miles.
Review the use of a distance scale. On this map, the scale is 1 inch = 60 miles.

Friday

1. Answer: Palo Duro Canyon State Park.
Point out that parks make use of the environment to provide recreational opportunities for people.

2. Answer: Good roads make travel easier.

Week Number 1 2 3 4 5 6 7 8 9 10 11 12 13 14 15 16 17 18 19 20 21 22 23 24 25 26 27 28 29 30 31 32 33 34 35 36

Name _____ Date _____

Directions → Use **Map G** to help you answer the questions.

Monday

1. Which state is southwest of Colorado?

2. What state borders Pennsylvania on the northeast?

Tuesday

1. Which two capital cities have their state names in their own names?

2. Which states touch the northwest corner of Missouri?

Wednesday

1. Which two states border both Canada and the Atlantic Ocean?

2. Which state has no common border with any other state or country? (Hint: It's surrounded by water.)

Thursday

1. Which two state names have the name of another state in them?

2. Which two states each border eight other states, including each other?

Friday

1. What do you call something found in nature that people can use?

2. What is an important natural resource in your state?

Week Number 1 2 3 4 5 6 7 8 9 10 11 12 13 14 15 16 17 18 19 20 21 22 23 24 25 26 27 28 29 30 31 32 33 34 35 36

www.svschoolsupply.com

© Steck-Vaughn Company

53

Unit 2: State

Weekly Geography Practice 3, SV 3432-0

Focus: The purpose of this week's questions is to acquaint the students with various states in the United States, and to reinforce the use of the intermediate directions.

Directions: Distribute **Map G** to assist students in answering the questions.

Monday

1. **Answer:** Arizona.
   Review the intermediate directions: northeast, southeast, southwest, and northwest.

2. **Answer:** New York.
   Some students might answer New Jersey, but the state bordering Pennsylvania on the true northeast is New York.

Tuesday

1. **Answer:** INDIANApolis and OKLAHOMA City.
   Many cities have state names in them, but these two are the only capital cities.

2. **Answer:** Kansas, Nebraska, and Iowa.

Wednesday

1. **Answer:** Maine and New Hampshire.

2. **Answer:** Hawaii.
   Review the definition of *island*. Because the Hawaiian Islands touch no other land, they cannot border on any other state.

Thursday

1. **Answer:** ArKANSAS and West VIRGINIA.

2. **Answer:** Tennessee and Missouri.
   Tennessee borders Missouri, Kentucky, Virginia, North Carolina, Georgia, Alabama, Mississippi, and Arkansas. Missouri borders Tennessee, Arkansas, Oklahoma, Kansas, Nebraska, Iowa, Illinois, and Kentucky.

Friday

1. **Answer:** natural resource.
   Point out some natural resources, such as water, minerals, oil, trees, and seafood.

2. **Answer:** Answers will vary.
   Point out that natural resources are usually very important for a state to have good jobs.

Week Number 1 2 3 4 5 6 7 8 9 10 11 12 13 14 15 16 17 18 19 20 21 22 23 24 25 26 27 28 29 30 31 32 33 34 35 36

www.svschoolsupply.com

© Steck-Vaughn Company

Unit 2: State

Weekly Geography Practice 3, SV 3432-0

Name _____   Date _____

Directions → Use **Map H** to help you answer the questions.

Monday

1. What three products are grown or raised near the town of Rose Hill?
   _____
   _____

2. What farm product is grown closest to the capital of North Carolina?
   _____
   _____

Tuesday

1. What farm animals are raised in the southern part of the state?
   _____
   _____

2. What farm animals are raised in the northern part of the state?
   _____
   _____

Wednesday

1. What state or body of water lies directly east of North Carolina? (You can use **Map G** to help you answer this question.)
   _____

2. Why do you think so many farm products are grown in the southeast part of the state?
   _____
   _____
   _____

Thursday

1. According to the map, what kind of farm animal is raised most in North Carolina?
   _____
   _____

2. Are more farm products grown in the eastern part or the western part of the state?
   _____
   _____

Friday

1. What are some important farm products in your state?
   _____
   _____
   _____

2. Would you like to live on a farm? Why or why not?
   _____
   _____
   _____

Week Number 1 2 3 4 5 6 7 8 9 10 11 12 13 14 15 16 17 18 19 20 21 22 23 24 25 26 27 28 29 30 31 32 33 34 35 36

www.svschoolsupply.com
© Steck-Vaughn Company

Unit 2: State
Weekly Geography Practice 3, SV 3432-0

Focus: The purpose of this week's questions is to acquaint students with the use of a resource map.
Directions: Distribute **Map H** to assist students in answering the questions.

Monday

1. **Answer:** poultry, hogs, and corn.
   You might need to define *poultry* for the students. Many times, farm crops are grown to feed farm animals, such as corn being used to feed hogs.

2. **Answer:** tobacco.
   Point out that tobacco is still a main farm crop in many southeastern states, even though tobacco is becoming increasingly unpopular with many people.

Tuesday

1. **Answer:** hogs and poultry.
   Point out that hogs are raised for meat production. The meat of hogs is called pork, and it includes such cuts as pork chops, bacon, and pigs' feet.

2. **Answer:** dairy cows and poultry.
   Point out that dairy cows are used for milk production more than meat production.

Wednesday

1. **Answer:** Atlantic Ocean.
   Suggest that the students use **Map G** for help in answering this question. North Carolina is a state in the area called the Atlantic Seaboard.

2. **Answer:** Answers will vary, but should indicate that the closeness of water allows for easier transportation of the farm goods.

Thursday

1. **Answer:** hogs.
   Four hog symbols are included while only three poultry symbols and two dairy cow symbols are on the map.

2. **Answer:** eastern part.
   Point out that the eastern part of the state contains coastal plains. The western part of the state is a part of the Blue Ridge Mountains. Ask the students if plains or mountains are better places to farm.

Friday

1. **Answer:** Answers will vary.
   Discuss with the students why farms are important.

2. **Answer:** Answers will vary.
   Ask the students what kind of chores or duties they might have on a farm.

Week Number 1 2 3 4 5 6 7 8 9 10 11 12 13 14 15 16 17 18 19 20 21 22 23 24 25 26 27 28 29 30 31 32 33 34 35 36

www.svschoolsupply.com
© Steck-Vaughn Company

56

Unit 2: State
Weekly Geography Practice 3, SV 3432-0

Name _____ Date _____

Directions → Use **Map I** to help you answer the questions.

Monday

1. What city is located in the highlands in northeast Pennsylvania?
_____
_____

2. What city is located on the plains in southeast Pennsylvania?
_____
_____

Tuesday

1. What is the name of this map? What does the map show?
_____
_____

2. What is the capital city of Pennsylvania?
_____
_____

Wednesday

1. Which city is on a lake: Erie, Harrisburg, or Philadelphia?
_____

2. Which city is located just west of the Appalachian Mountains: Harrisburg, Scranton, or Pittsburgh?
_____

Thursday

1. If this map showed the highest point in Pennsylvania, near which city would it be: Erie, Johnstown, or Harrisburg?
_____

2. What is the most common landform in Pennsylvania: mountains, highlands, or plains?
_____

Friday

1. Several of the cities in Pennsylvania are located beside rivers. Why do you think these cities were started beside rivers?
_____
_____

2. Would you rather live in the mountains or on the plains? Why?
_____
_____

Week Number   1 2 3 4 5 6 7 8 9 10 11 12 13 14 15 16 17 18 19 20 21 22 23 24 25 26 27 28 29 30 31 32 33 34 35 36

Focus: The purpose of this week's questions is to acquaint the students with the use of a relief (or elevation) map.

Directions: Distribute **Map I** to assist students in answering the questions.

Monday

1. **Answer:** Scranton.
   Review the use of the compass rose and the intermediate directions. Point out that this compass rose does not have *NE*, for example, but it has a small line pointing in that direction.

2. **Answer:** Philadelphia.
   Point out how the different landforms are represented on the map. You might also point out the relative elevations of each landform: mountains are the tallest, highlands are in the middle, and plains have the lowest elevation.

Tuesday

1. **Answer:** Pennsylvania Relief Map. It shows the relative heights of the landscape, including mountains, highlands, and plains. It also shows rivers, a lake, cities, and the capital city.

2. **Answer:** Harrisburg.
   Review the use of a map key. Ask the students how the capital city is represented on the map (star).

Wednesday

1. **Answer:** Erie.
   Ask the students what an important industry in this city might be (shipping).

2. **Answer:** Pittsburgh.

Thursday

1. **Answer:** Johnstown.
   Point out that Johnstown is located in the Appalachian Mountains. Mountains have a higher elevation than the surrounding area.

2. **Answer:** highlands.
   Some students might answer mountains; the two have similar proportions. Point out that this fact suggests Pennsylvania has a higher general altitude or elevation.

Friday

1. **Answer:** Answers will vary, but should suggest that the rivers allow easier movement of people and goods.

2. **Answer:** Answers will vary.
   Conduct a survey of the students and their choices, perhaps including a discussion of their reasons.

Week Number 1 2 3 4 5 6 7 8 9 10 11 12 13 14 15 16 17 18 19 20 21 22 23 24 25 26 27 28 29 30 31 32 33 34 35 36

www.svschoolsupply.com
© Steck-Vaughn Company

Unit 2: State
Weekly Geography Practice 3, SV 3432-0

Name _____ Date _____

Directions: Use **Map J** to help you answer the questions. You may also want to use **Map G**.

Monday

1. What is the name of your region?
_____
_____

2. What other states are in your region?
_____
_____
_____

Tuesday

1. How many regions are in the United States?
_____
_____

2. What other regions does your region touch?
_____
_____

Wednesday

1. What would be the easiest way to travel from the east side of your region to the west side?
_____
_____

2. How long do you think such a trip would take?
_____
_____

Thursday

1. How is your state similar to other states in your region?
_____
_____

2. How is your state different from other states in your region?
_____
_____

Friday

1. Why do you think people like to live in your region? Name three reasons.
_____
_____
_____

2. What are some fun vacation spots in your region?
_____
_____
_____

Week Number 1 2 3 4 5 6 7 8 9 10 11 12 13 14 15 16 17 18 19 20 21 22 23 24 25 26 27 28 29 30 31 32 33 34 35 36

www.svschoolsupply.com
© Steck-Vaughn Company

Unit 3: Region
Weekly Geography Practice 3, SV 3432-0

Focus: The purpose of this week's questions is to introduce the students to the larger geographical division, region, including similarities and differences among states in the same region.

Directions: Distribute **Map J** and **Map G** to assist students in answering the questions.

Monday

1. **Answer:** Answers will vary.
Help students find their state among the regions listed on the map. Students may be able to use **Map G** in conjunction with **Map J** to answer the questions themselves.

2. **Answer:** Answers will vary.
Help the students to name the other states in the region.

Tuesday

1. **Answer:** six regions.
Point out that there are several ways to break the United States into regions, and in those other divisions the regions may have different names. This map shows six regions in a standard division.

2. **Answer:** Answers will vary.
Point out that even though the regions have different names, they are still interrelated, just as the states in a region are interrelated.

Wednesday

1. **Answer:** Answers will vary.
The fastest mode of transportation would be by air, though many students will probably answer by car.

2. **Answer:** Answers will vary.
Travel across the Southeast Region, the Southwest Region, the North Central Region, and the Rocky Mountain Region would take several days by car. Travel time across the Pacific Region is harder to determine, because many parts of Alaska are hard to travel across, and Hawaii can only be reached by plane or ship.

Thursday

1. **Answer:** Answers will vary.
Point out that points of comparison-contrast could include population, climate, natural resources, jobs, landforms, etc.

2. **Answer:** Answers will vary.

Friday

1. **Answer:** Answers will vary.
Make a list of all the students' responses. Then, take a vote to determine the three most important reasons.

2. **Answer:** Answers will vary.

Week Number 1 2 3 4 5 6 7 8 9 10 11 12 13 14 15 16 17 18 19 20 21 22 23 24 25 26 27 28 29 30 31 32 33 34 35 36

Name _____ Date _____

Directions: Use **Map J** to help you answer the questions. You may also want to use **Map G**.

Monday

1. In which region is the state of Missouri?

   _____
   _____

2. In which region is the state of Tennessee?

   _____
   _____

Tuesday

1. Which state in the Pacific Region is the only state named for a United States President?

   _____
   _____

2. How many states are in the Rocky Mountain Region?

   _____
   _____

Wednesday

1. Could you travel from the Northeast Region to the Southwest Region by water? Through which two bodies of water would you travel?

   _____

2. Name two ways you could travel from California to Hawaii in the Pacific Region.

   _____
   _____

Thursday

1. What is the largest state in the Southwest Region?

   _____
   _____

2. What is the smallest state in the Northeast Region?

   _____
   _____

Friday

1. In which region would you be more likely to experience an earthquake: the Northeast Region or the Pacific Region?

   _____

2. In which region would you be more likely to experience a hurricane: the Southeast Region or the Rocky Mountain Region?

   _____

Focus: The purpose of this week's questions is to continue to develop the concept of the geographical region, including similarities and differences between states in different regions.

Directions: Distribute **Map J** and **Map G** to assist students in answering the questions.

Monday

1. Answer: North Central Region.
You might have students write lists of the states in each region.

2. Answer: Southeast Region.
The Southeast Region and the North Central Region each contain twelve states. They are the two regions with the most states in them.

Tuesday

1. Answer: Washington.
Washington entered the Union on November 11, 1889, as the 42nd state. Its highest point is Mount Rainier, at 14,410 feet (4,392 m).

2. Answer: six states.
Pacific Region: five states. Rocky Mountain Region: six states. Southwest Region: four states. North Central Region: twelve states. Southeast Region: twelve states. Northeast Region: eleven states.

Wednesday

1. Answer: Yes; Atlantic Ocean and Gulf of Mexico.
You might point out that people can travel from the Atlantic Ocean to the Pacific Ocean through the Panama Canal.

2. Answer: by airplane or ship.
You might ask the students to list all the different modes of transportation. What is one problem with travel by car? (Can't cross a large body of water.)

Thursday

1. Answer: Texas.
You might compare the size of Texas to the size of Alaska. In area, Alaska is over twice as big as Texas.

2. Answer: Rhode Island.
Point out the differing sizes of states. Rhode Island's area is 1,213 square miles (3,142 sq km). Alaska's area is 589,878 square miles (1,522,596 sq km). Alaska is almost 500 times bigger than Rhode Island.

Friday

1. Answer: Pacific Region.
California, for example, has earthquakes frequently. The Northeast Region very rarely experiences an earthquake.

2. Answer: Southeast Region.
Point out that hurricanes originate in the Atlantic Ocean and the Pacific Ocean. Most that strike the United States hit along the coasts of the Southeast Region or Texas. Hurricanes rarely make it to the Rocky Mountain Region.

Week Number 1 2 3 4 5 6 7 8 9 10 11 12 13 14 15 16 17 18 19 20 21 22 23 24 25 26 27 28 29 30 31 32 33 34 35 36

Name _____ Date _____

Directions ▸ Use **Map K** to help you answer the questions.

Monday

1. Which river flows through Washington, D.C.?
   _____
   _____

2. Which city is not located at the mouth of a river: Boston, New York, or Philadelphia?
   _____
   _____

Tuesday

1. Name three states whose capital cities are next to rivers.
   _____
   _____

2. Through which city does the Merrimack River flow?
   _____
   _____

Wednesday

1. Which river can you use to travel from Albany, New York, to New York City?
   _____
   _____

2. Which river can you use to travel from Trenton, New Jersey, to Philadelphia, Pennsylvania?
   _____
   _____

Thursday

1. Which river forms the border between two states and then runs south through two other states: Allegheny River, Connecticut River, or Kennebec River?
   _____

2. Which two states have no major rivers: Rhode Island and Maine; Delaware and Rhode Island; or New York and Vermont?
   _____
   _____

Friday

1. Which of these is a human-made feature: a river, a bridge, or an ocean?
   _____
   _____

2. Why do you think many cities of the Northeast Region grew up beside rivers?
   _____
   _____
   _____

Week Number 1 2 3 4 5 6 7 8 9 10 11 12 13 14 15 16 17 18 19 20 21 22 23 24 25 26 27 28 29 30 31 32 33 34 35 36

www.svschoolsupply.com

© Steck-Vaughn Company

63

Unit 3: Region
Weekly Geography Practice 3, SV 3432-0

Focus: The purpose of this week's questions is to acquaint the students with features on a physical map, particularly rivers.

Directions: Distribute **Map K** to assist students in answering the questions.

Monday

1. Answer: Potomac River.

2. Answer: Boston, Massachusetts.
Though Boston is not on a river, it does have a harbor on the Atlantic Ocean.

Tuesday

1. Answer: Students should name three of the following: Pennsylvania; New York; Connecticut; New Hampshire; Maine; New Jersey; Maryland.
Some students may say Washington, D.C. Washington, D.C., is next to a river (the Potomac), but it is not a state. It is a special district.

2. Answer: Concord, New Hampshire.

Wednesday

1. Answer: Hudson River.
The Hudson River was named after the English explorer Henry Hudson.

2. Answer: Delaware River.
Point out that it is usually easier to travel downstream than upstream on a river. The current flows downstream. You might point out the difference between a river's *source* (origin) and its *mouth* (where it feeds into a larger body of water).

Thursday

1. Answer: Connecticut River.
It forms the border between New Hampshire and Vermont, then flows south through Massachusetts and Connecticut.

2. Answer: Delaware and Rhode Island.

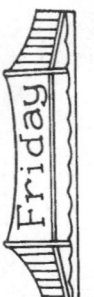

Friday

1. Answer: bridge.
Bridges often have to be built by humans in order to cross rivers.

2. Answer: Answers will vary.
Students should understand that the rivers were a way to travel from one place to another. There were no highways or railroads. Rivers gave the people a way to get the things they needed to live.

Name _____ Date _____

Directions → Use **Map L** to help you answer the questions.

Monday

1. For which region does this map show information?

2. Which state in this region is farthest south?

Tuesday

1. What landform covers the northeastern corner of Arizona and the southeastern corner of Utah?

2. In which natural region is Houston located?

Wednesday

1. Which direction would you go to travel from Albuquerque, New Mexico, to Austin, Texas?

2. Which state would probably take the longest to drive across from east to west?

Thursday

1. Name three mountain ranges shown on the map.

2. What two deserts can be found in Arizona?

Friday

1. Would you rather live in the mountains, in the desert, or on the plains? Why?

2. Which state shown on the map is most likely to have the most agricultural activity?

Week Number 1 2 3 4 5 6 7 8 9 10 11 12 13 14 15 16 17 18 19 20 21 22 23 24 25 26 27 28 29 30 31 32 33 34 35 36

Focus: The purpose of this week's questions is to introduce the students to the use of a landform map.

Directions: Distribute **Map L** to assist students in answering the questions.

Monday

1. Answer: Southwest Region.
Have the students identify the four states in the region.

2. Answer: Texas.
The highest point in Texas is Guadalupe Peak in the Guadalupe Mountains, with a height of 8,751 feet (2,667 m).

Tuesday

1. Answer: Colorado Plateau.
Define *plateau* as a land area having a generally level surface raised sharply above surrounding land on at least one side. A plateau is also sometimes called *tableland*.

2. Answer: coastal plains.
Houston is about 50 miles inland from the Gulf of Mexico. Coastal plains are usually flat or rolling, and they generally have fertile soil, good for farming.

Wednesday

1. Answer: southeast.

2. Answer: Texas.
From Beaumont on the eastern border to El Paso on the western tip is about 800 miles by highway.

Thursday

1. Answer: Rocky Mountains, Guadalupe Mountains, and Ouachita Mountains.
The Rocky Mountain range is the largest of these three.

2. Answer: the Sonoran Desert and the Painted Desert.
Define *desert* as an area that gets so little rain that it cannot support agriculture or any sizable population. Deserts are not necessarily hot; for this reason, Antarctica can be considered a desert.

Friday

1. Answer: Answers will vary.
Have the students discuss their choices and reasons.

2. Answer: Texas.
Farming is difficult in the desert and in the mountains. Most farming is done on the plains. Though Oklahoma has parts of the Great Plains and Central Plains passing through it, Texas has a larger area of plains in the north and northwest portions. Texas also has a large section of coastal plains.

Week Number 1 2 3 4 5 6 7 8 9 10 11 12 13 14 15 16 **17** 18 19 20 21 22 23 24 25 26 27 28 29 30 31 32 33 34 35 36

www.svschoolsupply.com
© Steck-Vaughn Company

Unit 3: Region
Weekly Geography Practice 3, SV 3432-0

Name _____ Date _____

Directions ➤ Use **Map M** to help you answer the questions.

Monday

1. In what state is Pikes Peak located?
_____
_____

2. In what state is Yellowstone National Park: Nevada, Wyoming, or Colorado?
_____
_____
_____

Tuesday

1. Utah has three national parks. What are they?
_____
_____
_____

2. What scenic place is just west of Utah's Great Salt Lake?
_____
_____

Wednesday

1. Suppose you want to travel from Mesa Verde National Park to Grand Teton National Park. In what direction would you have to travel?
_____

2. On your vacation to these parks, through which states would you have to travel?
_____
_____

Thursday

1. What kind of landform covers most of the states in this region?
_____
_____
_____

2. What is a group of mountains called?
_____
_____
_____

Friday

1. Hoover Dam is located near the bottom tip of what state? Why do people build dams?
_____
_____

2. What historical site is located in southeast Montana? Do you know what happened there?
_____
_____

Focus: The purpose of this week's questions is to introduce the students to the use of a regional map that highlights scenic or recreational areas.

Directions: Distribute **Map M** to assist students in answering the questions.

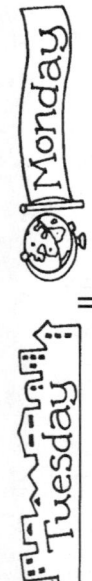

Monday

1. **Answer:** Colorado.
   You might point out that *Colorado* means "colored red" in Spanish.

2. **Answer:** Wyoming.
   Small parts of the park extend into Idaho and Montana.

Tuesday

1. **Answer:** Canyonlands National Park, Bryce Canyon National Park, and Zion National Park.
   Many national parks are located in the western half of the United States.

2. **Answer:** Bonneville Salt Flats.
   Many land-speed record attempts are held at the Bonneville Salt Flats.

Wednesday

1. **Answer:** north (or northwest).

2. **Answer:** Colorado and Wyoming.

Thursday

1. **Answer:** mountains.

2. **Answer:** a mountain range.

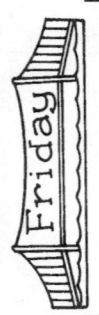

Friday

1. **Answer:** Nevada; Dams are built to hold back water to create lakes or to prevent flooding downstream. Dams are also used to generate electrical power.

2. **Answer:** Custer Battlefield. General George Custer's army troops were defeated by many Native American groups.

Week Number 1 2 3 4 5 6 7 8 9 10 11 12 13 14 15 16 17 18 19 20 21 22 23 24 25 26 27 28 29 30 31 32 33 34 35 36

www.svschoolsupply.com

© Steck-Vaughn Company

Unit 3: Region

Weekly Geography Practice 3, SV 3432-0

Name _____ Date _____

Directions: Use **Map N** to help you answer the questions.

Monday

1. Is Salem, Oregon, east or west of Los Angeles, California?
   _____
   _____

2. About how far apart are the Alaskan cities of Anchorage and Juneau?
   _____
   _____

Tuesday

1. What is the state capital of Washington?
   _____
   _____

2. If you wanted to travel to the northernmost state in the United States, where would you have to go? What region is this state in?
   _____

Wednesday

1. What direction would you travel from Anchorage, Alaska, to Honolulu, Hawaii? About how far apart are the two cities?
   _____

2. What is the distance from Los Angeles, California, to Honolulu, Hawaii? Which direction is Hawaii from California?
   _____

Thursday

1. What kind of landform has water all around it?
   _____
   _____

2. What is the difference between an island and a peninsula?
   _____
   _____

Friday

1. Mount St. Helens is a volcano in the Pacific Region. It is located in the state directly north of Oregon. What state is that?
   _____

2. How can volcanoes affect people?
   _____
   _____
   _____

Week Number 1 2 3 4 5 6 7 8 9 10 11 12 13 14 15 16 17 18 19 20 21 22 23 24 25 26 27 28 29 30 31 32 33 34 35 36

www.svschoolsupply.com

© Steck-Vaughn Company

Unit 3: Region
Weekly Geography Practice 3, SV 3432-0

Focus: The purpose of this week's questions is to continue to acquaint the students with regional maps, this time emphasizing distance and direction.

Directions: Distribute **Map N** to assist students in answering the questions.

Monday

1. Answer: west.

2. Answer: about 775 miles.
Stress to the students the fact that Alaska is a very large state.

Tuesday

1. Answer: Olympia.

2. Answer: Alaska; the Pacific Region.

Wednesday

1. Answer: southwest; about 3,500 miles.

2. Answer: about 3,000 miles; southwest.
Stress these great distances by pointing out that the distance from New York City to Los Angeles is only about 2,500 miles.

Thursday

1. Answer: an island.
Ask the students if they know the names of any islands.

2. Answer: An island is land with water all around it, but a peninsula is land not completely surrounded by water.
A peninsula may have water on three sides, such as the strip of land jutting out from the southwest part of Alaska.

Friday

1. Answer: Washington.

2. Answer: Volcanoes can explode, flattening vegetation and buildings nearby. They can send out dust and ash that coat the surrounding area, pollute the air, and block the sunlight. They can also send out rivers of lava that can destroy anything nearby.

Week Number 1 2 3 4 5 6 7 8 9 10 11 12 13 14 15 16 17 18 19 20 21 22 23 24 25 26 27 28 29 30 31 32 33 34 35 36

Name _____   Date _____

Directions: Use **Map G** to help you answer the questions.

Monday

1. In which country do you live?
_____
_____

2. What is the capital city of your country?
_____
_____

Tuesday

1. Between which two states is Washington, D.C., located? (Hint: You may use **Map K** to help you.)
_____
_____

2. Which state is farther south: Florida or Texas?
_____
_____

Wednesday

1. Which direction would you travel to get from North Dakota to North Carolina?
_____
_____

2. Which direction would you travel to get from Oklahoma to Oregon?
_____
_____

Thursday

1. What other countries does your country touch?
_____
_____

2. Which state is located in the middle of the country but borders only four other states?
_____
_____

Friday

1. If you could pass one law to improve your country, what would your law be?
_____
_____

2. Name three reasons your country is a good place to live.
_____
_____

Week Number 1 2 3 4 5 6 7 8 9 10 11 12 13 14 15 16 17 18 19 20 21 22 23 24 25 26 27 28 29 30 31 32 33 34 35 36

Focus: The purpose of this week's questions is to acquaint the students with the larger political division of the country, particularly the United States, and some of the basic concepts associated with their country.

Directions: Distribute **Map G** to assist students in answering the questions.

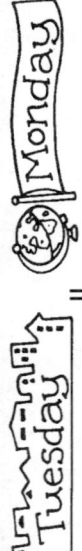

Monday

1. **Answer:** The United States of America.
   The assumption is made that this book is being used in the United States.

2. **Answer:** Washington, D.C.
   The assumption is made that this book is being used in the United States.

Tuesday

1. **Answer:** Maryland and Virginia.
   Explain to the students that the city of Washington is not a state, nor is it part of any state. It is a special district: the District of Columbia.

2. **Answer:** Florida.
   You might want to review the cardinal directions.

Wednesday

1. **Answer:** southeast.
   You might want to review the intermediate directions.

2. **Answer:** northwest.

Thursday

1. **Answer:** Canada and Mexico.
   The assumption is made that this book is being used in the United States.

2. **Answer:** Kansas.
   Kansas is roughly the center of the 48 contiguous (touching) states.

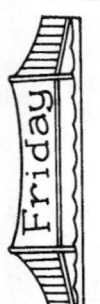

Friday

1. **Answer:** Answers will vary.
   Make a list of all the students' responses. Then, take a vote to see which proposed law the students consider the most reasonable. You might point out that all laws in the United States must abide by the *Constitution*.

2. **Answer:** Answers will vary.
   Make a list of all the students' responses. Then, take a vote to see which three reasons the students consider the most important.

Week Number 1 2 3 4 5 6 7 8 9 10 11 12 13 14 15 16 17 18 19 20 21 22 23 24 25 26 27 28 29 30 31 32 33 34 35 36

www.svschoolsupply.com
© Steck-Vaughn Company

Unit 4: Country
Weekly Geography Practice 3, SV 3432-0

Name _____ Date _____

Directions: Use **Map G** to help you answer the questions.

Monday

1. Which state is farther north: South Dakota or North Carolina?
_____
_____

2. The Four Corners is the only place where you can stand in four states at the same time. Which four states meet at the Four Corners?
_____

Tuesday

1. Which of these is a physical feature: a bridge, a plain, or a road?
_____
_____

2. Which of these is a human-made feature: a plateau, a river, or a highway?
_____

Wednesday

1. Some rivers and streams have shallow places where people can cross by walking, riding, or driving. What are these shallow places called?
_____

2. If a river does not have a shallow place, how can people cross the river?
_____

Thursday

1. If you were flying over an isthmus, would you be over land or water? What is an isthmus?
_____

2. If you were flying over a mesa, would you be over land or water? What is a mesa?
_____

Friday

1. What kind of landform would be the best place to have a farm: a mountain, a hill, or a plain?
_____
_____

2. In which part of the United States would you probably find more farmland: the northeastern part or the middle of the country?
_____
_____

Week Number 1 2 3 4 5 6 7 8 9 10 11 12 13 14 15 16 17 18 19 20 21 22 23 24 25 26 27 28 29 30 31 32 33 34 35 36

www.svschoolsupply.com
© Steck-Vaughn Company

Unit 4: Country
Weekly Geography Practice 3, SV 3432-0

Focus: The purpose of this week's questions is to continue to acquaint the students with the concept of country, particularly the United States, and some of the basic landforms associated with their country.

Directions: Distribute **Map G** to assist students in answering the questions.

Monday

1. Answer: South Dakota.

2. Answer: Utah, Arizona, New Mexico, and Colorado.
Ask if any students have been to the Four Corners. Point it out on a map.

Tuesday

1. Answer: a plain.
You might define *plain* as a large area of level or rolling treeless land. Point out that much of the center of the United States is made up of plains, especially the Great Plains.

2. Answer: a highway.
You might define *plateau* as a mostly level surface raised sharply above the surrounding land on at least one side; a plateau is sometimes called *tableland*.

Wednesday

1. Answer: fords.
Point out that it can be dangerous to cross water in a car or by walking if you do not know how deep the water is.

2. Answer: by bridge, by boat, or by ferry.
Students might suggest other ways; these three are the most common.

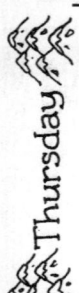

Thursday

1. Answer: land; a narrow strip of land connecting two larger land areas.

2. Answer: land; a flat-topped hill or plateau with steep sides.

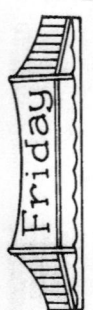

Friday

1. Answer: a plain.
Level or rolling plains are the best kind of farmland.

2. Answer: the middle part.
The middle part of the United States contains the Great Plains, which contain good farmland.

Week Number 1 2 3 4 5 6 7 8 9 10 11 12 13 14 15 16 17 18 19 20 21 22 23 24 25 26 27 28 29 30 31 32 33 34 35 36

Name _____ Date _____

Directions: Use **Map O** to help you answer the questions. You might also want to use **Map G** and **Map J**.

Monday

1. What kind of weather is Colorado having?

2. What kind of weather is Hawaii having?

Tuesday

1. Which three states that touch the Pacific Ocean are having rainy weather?

2. Which two states that touch the Atlantic Ocean are having partly sunny weather?

Wednesday

1. If you drove from Texas to California, what kind of weather would you expect to find?

2. If you drove from North Dakota to Michigan, what kind of weather would you expect to find?

Thursday

1. What kind of weather is the Northeast Region having?

2. What is climate?

Friday

1. Are the snowy parts of the country mostly in the north or in the south?

2. If you wanted to go on vacation where the weather is sunny, which states might you choose to visit?

Week Number 1 2 3 4 5 6 7 8 9 10 11 12 13 14 15 16 17 18 19 20 21 22 23 24 25 26 27 28 29 30 31 32 33 34 35 36

www.svschoolsupply.com

© Steck-Vaughn Company

75

Unit 4: Country

Weekly Geography Practice 3, SV 3432-0

Focus: The purpose of this week's questions is to acquaint the students with the use of a weather map.

Directions: Distribute **Map O** to assist students in answering the questions. You might also suggest they refer to **Map G** and **Map J** for further assistance.

Monday

1. **Answer:** rainy.
   Point out the different symbols used in the map key.

2. **Answer:** partly sunny.

Tuesday

1. **Answer:** California, Oregon, and Washington.

2. **Answer:** Georgia and Florida.

Wednesday

1. **Answer:** sunny.

2. **Answer:** snowy.
   Point out that travelers need to pay attention to the weather so that they do not encounter bad weather.

Thursday

1. **Answer:** snowy.
   Review the locations of the different regions.

2. **Answer:** the kind of weather a place has over a long time.

Friday

1. **Answer:** north.

2. **Answer:** California, Arizona, New Mexico, Texas, or Alaska.

Name _____   Date _____

Directions: Use **Map P** to help you answer the questions.

Monday

1. What two cities are located on the line of latitude labeled 30° N?
   _____
   _____

2. What two cities are located on the line of longitude labeled 80° W?
   _____
   _____

Tuesday

1. What city is located at 40° N, 90° W?
   _____

2. What are the coordinates for the city of Reno? (Hint: Give the line of latitude first.)
   _____
   _____

Wednesday

1. In which direction would you travel to go from 40° N latitude to 30° N latitude?
   _____
   _____

2. In which direction would you travel to go from 90° W longitude to 105° W longitude?
   _____
   _____

Thursday

1. If you were following 30° N latitude across the United States, which would be the only three states you would cross?
   _____
   _____

2. What five cities on the map are located on 40° N latitude?
   _____
   _____
   _____

Friday

1. Would an airplane flying at 30° N, 120° W, be over land or water? Why might a pilot want to know the location of the airplane?
   _____
   _____

2. Would the winter weather be colder at 45° N latitude or 25° N latitude? Why?
   _____
   _____

Week Number 1 2 3 4 5 6 7 8 9 10 11 12 13 14 15 16 17 18 19 20 21 22 23 24 25 26 27 28 29 30 31 32 33 34 35 36

www.svschoolsupply.com
© Steck-Vaughn Company

Unit 4: Country
Weekly Geography Practice 3, SV 3432-0

Focus: The purpose of this week's questions is to introduce the students to the concepts of longitude and latitude.

Directions: Distribute **Map P** to assist students in answering the questions.

Monday

1. Answer: Houston and New Orleans.
Point out to the students that this line of latitude is called 30° N all the way around Earth. If you have a globe handy, you can demonstrate this fact. Lines of latitude run east and west, and they only go up to 90° N or S. Also point out to the students the small circle that stands for *degrees*.

2. Answer: Pittsburgh and West Palm Beach.
Lines of longitude run north and south. They do not have the same number all the way around the world. Lines of longitude (other than 90°) have a certain number only from pole to pole. For example, 100° W becomes 80° E on the other side of the world. Use the globe to demonstrate.

Tuesday

1. Answer: Springfield.
Locations of places are indicated by coordinates. Coordinates are the numbers that name where lines of latitude and longitude cross.

2. Answer: 40° N, 120° W.
Demonstrate the proper way of giving coordinates.

Wednesday

1. Answer: south.
Point out the movement to the students.

2. Answer: west.

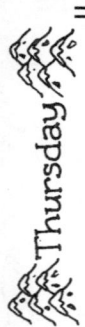

Thursday

1. Answer: Florida, Louisiana, and Texas.

2. Answer: Reno, Boulder, Springfield, Pittsburgh, and Philadelphia.

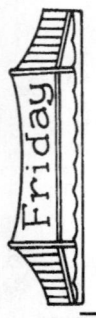

Friday

1. Answer: over the water; Answers will vary.
A pilot would not want to land in the water, for example.

2. Answer: 45° N.
The larger the number of the line of latitude, the farther that line is from the equator, so the colder the weather would be.

Name _____ Date _____

Directions: Use **Map G** and **Map P** to help you answer the questions.

Monday

1. Between which two lines of longitude is Lake Ontario located?
   _____
   _____

2. Which state capital in the northeast part of the United States is located east of 70° W longitude?
   _____
   _____

Tuesday

1. Which city is the capital of Montana?
   _____
   _____

2. Which state capital in the northwest part of the United States is located at 45° N latitude?
   _____
   _____

Wednesday

1. In which direction would you travel going from 35° N latitude to 45° N latitude?
   _____
   _____

2. In which direction would you travel going from 115° W longitude to 105° W longitude?
   _____
   _____

Thursday

1. Name the states in the United States that have directions in their names.
   _____
   _____

2. Which of these pairs is made up of two states that border Tennessee: Indiana and Iowa; Missouri and Virginia; Arkansas and South Carolina?
   _____
   _____

Friday

1. If you could visit any landform in the United States, such as a mountain or a river, which would you choose to visit? Why?
   _____
   _____

2. Much of the natural beauty of this country is hurt by litter. What can be done to prevent littering?
   _____
   _____

Week Number 1 2 3 4 5 6 7 8 9 10 11 12 13 14 15 16 17 18 19 20 21 22 23 24 25 26 27 28 29 30 31 32 33 34 35 36

www.svschoolsupply.com

© Steck-Vaughn Company

79

Unit 4: Country

Weekly Geography Practice 3, SV 3432-0

Focus: The purpose of this week's questions is to continue the study of the United States and the use of latitude and longitude.

Directions: Distribute **Map G** and **Map P** to assist students in answering the questions.

Monday

1. Answer: 75° W and 80° W.

2. Answer: Augusta, Maine.

Tuesday

1. Answer: Helena, Montana.

2. Answer: Salem, Oregon.

Wednesday

1. Answer: north.

2. Answer: east.

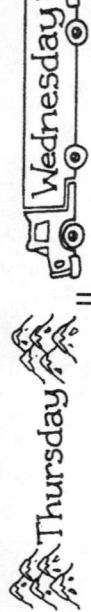

Thursday

1. Answer: North Dakota, South Dakota, West Virginia, North Carolina, and South Carolina.

2. Answer: Missouri and Virginia.

Friday

1. Answer: Answers will vary. Have the students discuss the reasons for their preferences.

2. Answer: Answers will vary. Suggested answers might include more public awareness or education about littering, stricter laws, etc.

Week Number 1 2 3 4 5 6 7 8 9 10 11 12 13 14 15 16 17 18 19 20 21 22 23 24 25 26 27 28 29 30 31 32 33 34 35 36

www.svschoolsupply.com
© Steck-Vaughn Company

80

Unit 4: Country
Weekly Geography Practice 3, SV 3432-0

Name _____ Date _____

Directions: Use **Maps G, J, N,** and **P** to help you answer the questions.

Monday

1. In which of the 50 states can you find the northernmost city in the United States?

2. Which city is the capital of the state that borders Kansas on the west?

Tuesday

1. The first letters of the names of the Great Lakes spell a word. What is the word?

2. What state capital in the Southeast Region is named for a President of the United States?

Wednesday

1. What three very large bodies of water touch the United States on the east, south, and west sides?

2. Name four kinds of transportation you could use to get from the east coast to the west coast of the United States.

Thursday

1. What is the highest kind of landform called?

2. What kind of landform lies between hills or mountains?

Friday

1. In which state would you be more likely to experience a tornado: Rhode Island or Kansas?

2. What kind of job might you have if you lived by the ocean?

Week Number 1 2 3 4 5 6 7 8 9 10 11 12 13 14 15 16 17 18 19 20 21 22 23 24 25 26 27 28 29 30 31 32 33 34 35 36

www.svschoolsupply.com
© Steck-Vaughn Company

81

Unit 4: Country
Weekly Geography Practice 3, SV 3432-0

Focus: The purpose of this week's questions is to continue to acquaint the students with the concept of country and some of the characteristics and landforms associated with the United States.

Directions: Distribute **Maps G**, **J**, **N**, and **P** to assist students in answering the questions.

Monday

1. Answer: Alaska.
The city is Barrow.

2. Answer: Denver, Colorado.

Tuesday

1. Answer: HOMES.
The Great Lakes are Huron, Ontario, Michigan, Erie, and Superior.

2. Answer: Jackson, Mississippi.
It is named after Andrew Jackson.

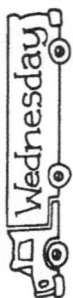

Wednesday

1. Answer: Atlantic Ocean, Gulf of Mexico, and Pacific Ocean.

2. Answer: Answers will vary, but could include by air, by rail, by car, by water (around South America or through the Panama Canal), by walking, by horseback, on a bicycle or motorcycle, on a skateboard, etc.

Thursday

1. Answer: mountains.
Point out that the United States has two major mountain ranges, the Appalachian Mountains in the east and the Rocky Mountains in the west.

2. Answer: valleys.
Some of the more famous valleys in the United States are the Shenandoah Valley in Virginia and Death Valley and Napa Valley in California.

Friday

1. Answer: Kansas.
Many more tornadoes strike the plains area than the northeast. The central plains area is sometimes called Tornado Alley. Texas has the greatest frequency of tornado activity.

2. Answer: Answers will vary.
Almost any kind of job suggestion is acceptable, but you might point out that an ocean location would provide many jobs in fishing and shipping.

Week Number 1 2 3 4 5 6 7 8 9 10 11 12 13 14 15 16 17 18 19 20 21 22 23 24 25 26 27 28 29 30 31 32 33 34 35 36

www.svschoolsupply.com

© Steck-Vaughn Company

Unit 4: Country
Weekly Geography Practice 3, SV 3432-0

Name _____ Date _____

Directions: Use **Map T** to help you answer the questions.

Monday

1. What are the seven largest land areas of the Earth called?
_____
_____
_____

2. On which continent do you live?
_____
_____
_____

Tuesday

1. What are the names of the seven continents?
_____
_____
_____

2. Which continent is the farthest south?
_____
_____
_____

Wednesday

1. If you wanted to travel directly from Africa to South America, which direction would you go?
_____
_____

2. If you wanted to travel directly from Africa to Europe, which direction would you go?
_____
_____

Thursday

1. Which continent is the largest in area?
_____
_____

2. Which continent is the smallest in area?
_____
_____

Friday

1. Many different kinds of people live in the world. Why is it good to know about other kinds of people?
_____
_____

2. If you could visit any continent, which would you choose? Why?
_____
_____

Focus: The purpose of this week's questions is to introduce the students to the concept of the world, particularly continents.

Directions: Distribute **Map T** to assist students in answering the questions.

Monday

1. **Answer:** continents.

2. **Answer:** North America.

Tuesday

1. **Answer:** North America, South America, Europe, Africa, Asia, Australia, and Antarctica.

2. **Answer:** Antarctica.

Wednesday

1. **Answer:** west.
Point out to the students that it is possible to get to South America from Africa by heading east, but this route is not as direct. Demonstrate with a globe. Columbus, for example, believed he could reach the East (India) by sailing west from Europe.

2. **Answer:** north.
Point out to the students that it is possible to get to Europe from Africa by heading south, but this route is not as direct. Demonstrate with a globe.

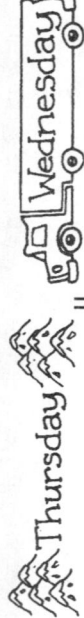

Thursday

1. **Answer:** Asia.
Asia has over 17 million square miles in area.

2. **Answer:** Australia.
Australia has just under 3 million square miles in area.

Friday

1. **Answer:** Answers will vary.
Knowing about other people and other cultures makes people more aware of the different needs and beliefs of their fellow human beings.

2. **Answer:** Answers will vary.
Take a vote to see which continent is the most popular choice.

Name _____ Date _____

Directions → Use **Map Q** to help you answer the questions.

Monday

1. Which three countries make up most of North America?

_____
_____
_____

2. Which country in North America is the farthest south?

_____
_____
_____

Tuesday

1. Which of these cities is the capital of Canada: Ottawa, Montreal, or Toronto?

_____
_____
_____

2. Which city is the capital of Mexico?

_____
_____
_____

Wednesday

1. If you traveled from Mexico to Canada, which direction would you go?

_____
_____
_____

2. If you wanted to travel from Ottawa to Mexico City, which direction would you go?

_____
_____
_____

Thursday

1. Which country touches the United States on the south?

_____
_____
_____

2. Which country touches the United States on the north?

_____
_____
_____

Friday

1. Much of Canada is cold. Much of Mexico is hot. In which place would you rather live? Why?

_____
_____
_____

2. In far northern places like Canada, the winter weather is very cold. What would your life be like in such a cold climate?

_____
_____
_____

Week Number 1 2 3 4 5 6 7 8 9 10 11 12 13 14 15 16 17 18 19 20 21 22 23 24 25 26 27 28 29 30 31 32 33 34 35 36

www.svschoolsupply.com

© Steck-Vaughn Company

85

Unit 5: World
Weekly Geography Practice 3, SV 3432-0

Focus: The purpose of this week's questions is to introduce the students to the continent of North America.

Directions: Distribute **Map Q** to assist students in answering the questions.

Monday

1. Answer: Canada, the United States, and Mexico.
North America is the third largest of the continents, with about 9.4 million square miles in area.

2. Answer: Mexico.

Tuesday

1. Answer: Ottawa.

2. Answer: Mexico City.
Mexico City is the largest city in population on the North American continent.

Wednesday

1. Answer: north.

2. Answer: southwest.

Thursday

1. Answer: Mexico.
Mexico is the smallest of the three North American countries.

2. Answer: Canada.
Canada is the largest of the three North American countries.

Friday

1. Answer: Answers will vary.
Discuss the advantages or disadvantages of living in a hot or cold place.

2. Answer: Answers will vary.
You might also mention the lack of sunshine in northern areas. The Sun does not rise above the horizon for many days. Growing seasons are usually quite short.

Name _____ Date _____

Directions: Use **a globe** to help you answer the questions.

Monday

1. What is a globe?

_____

_____

2. What shape is the globe? What shape is Earth?

_____

_____

Tuesday

1. What is located at the very top of the globe?

_____

_____

2. What is located at the very bottom of the globe?

_____

_____

Wednesday

1. Could you reach Asia from Australia by traveling south?

_____

2. Could you reach Africa from South America by traveling west?

_____

Thursday

1. What is a half of the Earth called?

_____

_____

2. In which two hemispheres do you live?

_____

_____

Friday

1. If you could live anywhere in the world, where would you want to live? Why?

_____

2. People live all over the world. They live in cold places, hot places, dry places, and wet places. Why do you think people choose to live where they do?

_____

Week Number 1 2 3 4 5 6 7 8 9 10 11 12 13 14 15 16 17 18 19 20 21 22 23 24 25 26 27 28 29 30 31 32 33 34 35 36

www.svschoolsupply.com

© Steck-Vaughn Company

87

Unit 5: World

Weekly Geography Practice 3, SV 3432-0

Focus: The purpose of this week's questions is to introduce the students to the use of a globe to locate places.

Directions: Have **a globe** available to assist students in answering the questions.

Monday

1. **Answer:** a model of the Earth.

2. **Answer:** round, or more correctly, spherical; round or spherical.

Tuesday

1. **Answer:** North Pole.
You might point out that from the North Pole, all directions are south. Similarly, from the South Pole, all directions are north.

2. **Answer:** South Pole.
You might point out that there is a multinational research facility at the South Pole.

Wednesday

1. **Answer:** yes.
You might demonstrate this fact by following a line of longitude over the pole.

2. **Answer:** yes.
You might demonstrate this fact by following a line of latitude around the world.

Thursday

1. **Answer:** hemisphere.
Point out to the students that the Equator divides the Earth into the Northern Hemisphere and the Southern Hemisphere. The Prime Meridian divides the Earth into the Eastern Hemisphere and the Western Hemisphere.

2. **Answer:** North America is in the Northern Hemisphere and the Western Hemisphere.

Friday

1. **Answer:** Answers will vary.
You might send this question home as a homework assignment.

2. **Answer:** Answers will vary.
You might lead a discussion about why people live where they do. Some do so because their ancestors lived there; others move to a place to get a job; others do not move because they lack means of transportation.

Week Number 1 2 3 4 5 6 7 8 9 10 11 12 13 14 15 16 17 18 19 20 21 22 23 24 25 26 27 28 29 30 31 32 33 34 35 36

Name _____ Date _____

Directions: Use **Map T** and **a globe** to help you answer the questions.

Monday

1. What are the largest bodies of water on the Earth called?

2. What are the names of the four oceans?

Tuesday

1. Which ocean touches the United States on the east?

2. Which ocean touches the United States on the west?

Wednesday

1. If you wanted to sail directly from Africa to North America, which ocean would you cross?

2. If you wanted to sail directly from Australia to Africa, which ocean would you cross?

Thursday

1. Which places are the farthest north and the farthest south you can go on the Earth?

2. Which ocean is near the North Pole?

Friday

1. Would you want to swim in the ocean near Antarctica? Why or why not?

2. Why are the oceans important to people?

Week Number 1 2 3 4 5 6 7 8 9 10 11 12 13 14 15 16 17 18 19 20 21 22 23 24 25 26 27 28 29 30 31 32 33 34 35 36

www.svschoolsupply.com

89

© Steck-Vaughn Company

Unit 5: World

Weekly Geography Practice 3, SV 3432-0

Focus: The purpose of this week's questions is to continue to acquaint the students with the concept of the world, particularly oceans.

Directions: Distribute **Map T** to assist students in answering the questions. Have **a globe** handy, too.

Monday

1. Answer: oceans.
Point out that there are many kinds of bodies of water: oceans, seas, gulfs, bays, lakes, ponds, etc. Oceans are the largest of these.

2. Answer: Atlantic Ocean, Pacific Ocean, Arctic Ocean, and Indian Ocean.
The Pacific Ocean is the largest, covering about 70 million square miles. It is more than twice as large as the Atlantic Ocean, which is the second largest.

Tuesday

1. Answer: Atlantic Ocean.

2. Answer: Pacific Ocean.

Wednesday

1. Answer: Atlantic Ocean.

2. Answer: Indian Ocean.
The Indian Ocean is the third largest ocean, just a bit smaller than the Atlantic Ocean.

Thursday

1. Answer: North Pole and South Pole.
Demonstrate the location of the two poles on a globe.

2. Answer: Arctic Ocean.
The Arctic Ocean and North Atlantic Ocean are very cold, usually with icebergs. The *Titanic* struck an iceberg in the North Atlantic Ocean.

Friday

1. Answer: Answers will vary, but should probably indicate the students would not want to swim there because the water is very cold.

2. Answer: Answers will vary.
The students should be aware of the oceans' role in weather and the water cycle, their role in supplying food, and their role in transportation and recreation.

Week Number 1 2 3 4 5 6 7 8 9 10 11 12 13 14 15 16 17 18 19 20 21 22 23 24 25 26 27 28 29 30 31 32 33 34 35 36

Name _____ Date _____

Directions: Use **Map R** and **a globe** to help you answer the questions.

Monday

1. What is the name of the imaginary line around the Earth halfway between the North Pole and the South Pole?

2. Which directions does this line run? Is it a line of latitude or longitude? What is its number label?

Tuesday

1. What is the name of the imaginary line that runs north and south from the North Pole to the South Pole at 0°?

2. Near which continent does the Equator meet the Prime Meridian? Is the meeting point on land or on water?

Wednesday

1. Which direction would you travel to get directly from Brisbane to Cairo? Would you cross the Equator? Would you cross the Prime Meridian?

2. Which two cities are located on latitude line 30° N? Which one is located at 30° N, 30° E?

Thursday

1. What city is located at 60° N, 150° E? On what continent is this city?

2. Which continent is located completely between 30° W and 90° W?

Friday

1. What if you could skate around the world on latitude 30° N? What do you think you would see along the way?

2. Write a short story about your skating trip around the world. Include some pictures with your story.

Week Number 1 2 3 4 5 6 7 8 9 10 11 12 13 14 15 16 17 18 19 20 21 22 23 24 25 26 27 28 29 30 31 32 33 34 35 36

www.svschoolsupply.com

91

© Steck-Vaughn Company

Unit 5: World
Weekly Geography Practice 3, SV 3432-0

Focus: The purpose of this week's questions is to reinforce the concepts of latitude and longitude, this time in relation to locations across the world.

Directions: Distribute **Map R** to assist students in answering the questions. Have **a globe** handy, too.

Monday

1. Answer: Equator.
Equator comes from Latin and means "equalizer."

2. Answer: east and west; latitude; 0°.

Tuesday

1. Answer: Prime Meridian.
The Prime Meridian runs through the site of the Royal Observatory at Greenwich, England. *Meridian* is from Middle English, meaning "at noon."

2. Answer: Africa; on water.
Point out the location to the students on a map or globe.

Wednesday

1. Answer: northwest; Equator: yes; Prime Meridian: no.

2. Answer: New Orleans and Cairo; Cairo.
Have the students run their finger along this line of latitude.

Thursday

1. Answer: Magadan; Asia.

2. Answer: South America.

Friday

1. Answer: Answers will vary.
Have the students volunteer their ideas. Have them check to see which countries are located along the line.

2. Answer: Answers will vary.
This question is intended as a writing extension, and it should be assigned as homework. The students can make up a little book with their story and drawings.

Week Number 1 2 3 4 5 6 7 8 9 10 11 12 13 14 15 16 17 18 19 20 21 22 23 24 25 26 27 28 29 30 31 32 33 34 35 36

www.svschoolsupply.com
© Steck-Vaughn Company

Unit 5: World
Weekly Geography Practice 3, SV 3432-0

Name _____ Date _____

Directions: Use **Map R** and **Map T** to help you answer the questions.

Monday

1. On which continent is the city of Shanghai located?

2. On which continent is the city of Brisbane located?

Tuesday

1. What city is located at 60° N, 30° E?

2. Which two oceans touch both North America and Asia?

Wednesday

1. If you travel directly from the south coast of Asia to the north coast of Antarctica, which direction would you go?

2. If you traveled directly from the east coast of South America to the west coast of North America, which direction would you go?

Thursday

1. In which two hemispheres is the continent of Australia?

2. In how many hemispheres is the continent of Africa?

Friday

1. Earth Day is held each April. Earth Day celebrates the Earth. Make an Earth Day poster with a clever slogan to help stop pollution.

2. Why is the study of geography important?

Week Number 1 2 3 4 5 6 7 8 9 10 11 12 13 14 15 16 17 18 19 20 21 22 23 24 25 26 27 28 29 30 31 32 33 34 35 36

www.svschoolsupply.com

© Steck-Vaughn Company

93

Unit 5: World
Weekly Geography Practice 3, SV 3432-0

Focus: The purpose of this week's questions is to continue to acquaint the students with the concept of the world.

Directions: Distribute **Map R** and **Map T** to assist students in answering the questions.

Monday

1. Answer: Asia.

2. Answer: Australia.
Point out to the students that Australia is the only continent that contains only one country. The continent and the country have the same name.

Tuesday

1. Answer: St. Petersburg.
Review the use of coordinates to locate a place.

2. Answer: Pacific Ocean and Arctic Ocean.

Wednesday

1. Answer: south to southwest.

2. Answer: northwest.

Thursday

1. Answer: Eastern Hemisphere and Southern Hemisphere.

2. Answer: all four hemispheres.

Friday

1. Answer: Answers will vary.
This is best assigned as homework. Display the completed posters in the classroom.

2. Answer: The study of geography helps us to know more about the world we live in.

Name _____ Date _____

Directions: Use **Map S** to help you answer the questions.

Monday

1. What are the large bodies called that orbit the Sun?
   _____
   _____

2. What is the name of the planet you live on?
   _____
   _____

Tuesday

1. Which planet is closest to the Sun?
   _____
   _____

2. Which planet is farthest from the Sun?
   _____
   _____

Wednesday

1. If you traveled from Earth to Neptune, would you be traveling toward the Sun or away from the Sun?
   _____
   _____

2. How long do you think it would take to travel from Earth to Neptune?
   _____
   _____

Thursday

1. Which is the largest planet?
   _____
   _____

2. Which planet has many rings?
   _____
   _____

Friday

1. Why can't people live on Mercury?
   _____
   _____
   _____

2. Would you like to travel into space? Why or why not?
   _____
   _____
   _____

Week Number 1 2 3 4 5 6 7 8 9 10 11 12 13 14 15 16 17 18 19 20 21 22 23 24 25 26 27 28 29 30 31 32 33 34 35 36

www.svschoolsupply.com

© Steck-Vaughn Company

Unit 5: World

Weekly Geography Practice 3, SV 3432-0

Focus: The purpose of this week's questions is to acquaint the students with the concept of the solar system and Earth's place in it.

Directions: Distribute **Map S** to assist students in answering the questions.

Monday

1. **Answer:** planets.
   The word *planet* comes from Greek and Latin, and it means "wanderer."

2. **Answer:** Earth.

Tuesday

1. **Answer:** Mercury.

2. **Answer:** usually Pluto.
   At times, Pluto moves inside the orbit of Neptune.

Wednesday

1. **Answer:** away from the Sun.

2. **Answer:** many, many years.
   Neptune is more than 4 billion kilometers (about 2.6 billion miles) from Earth.

Thursday

1. **Answer:** Jupiter.

2. **Answer:** Saturn.
   Saturn is the best-known planet with rings, though several of the outer planets have rings. Saturn's rings are the most prominent.

Friday

1. **Answer:** It is too hot on the side toward the Sun and too cold on the side away from the Sun.
   Earth is the only planet known to have a moderate climate.

2. **Answer:** Answers will vary.

Week Number 1 2 3 4 5 6 7 8 9 10 11 12 13 14 15 16 17 18 19 20 21 22 23 24 25 26 27 28 29 30 31 32 33 34 35 36